"Having known Eduardo as a committed citizen around climate change, I'm excited to now see him bring his unique thinking and experience to bear on one of humanity's most pressing concerns. This book boldly seeks to create a unifying vision of care for the environment, inviting various communities to come together around the single issue of climate change. *A Climate of Desire* is an admirable project for an increasingly divided world, whose future absolutely depends on our reconciliation and collaboration."
—Shad, Alternative Hip Hop Artist

"Climate change is an existential threat to humanity. As such, technical fixes alone are inadequate—it requires a deep re-imagination of the true meaning of life. In his book, Eduardo elevates climate change to the level at which it needs to be addressed: humanity's core purpose. His deft juxtaposition of science and religion help us see where solutions might lie. This book is a must-read for anyone who wants to explore these issues at their roots. You will see the world in new ways—and your role in it."
—Coro Strandberg, Associate, Canadian Business for Social Responsibility

"Sasso paints with big brushes on a large canvas, applying insights from philosophy, science, and history to some of the most important questions of today—from climate change and ecological destruction to gender equality and the role of faith. These are weighty (even controversial) subjects, but Sasso's skilful storytelling effortlessly weaves together facts and anecdotes, making this book a pleasure to read. So if you're looking for real insights that arise at the nexus of disciplines and through the process of dialogue, here's a book that brings fresh thinking to the great dilemmas of humanity, faith, and ecology."
—Wayne Visser, Fellow, Institute for Sustainability Leadership, University of Cambridge

"This must be the first book that dares to weave together sex, Christianity, and climate change. But Sasso, a 'recovering engineer', shows how closely related they actually are. Provocative, readable, and sustained by robust reflection, it is hard to imagine how this book will not widen and deepen our vision for this world's future."
—Jeremy Begbie, Distinguished Professor, Duke University

"The moral call for climate action requires all of us to find a way of contributing. Weaving together compelling stories, science, biblical studies, and philosophy, Sasso eloquently lights sparks and connects dots that will help new people find their place in this global movement."
—Christine Boyle, Founder, Fossil Free Faith

"Wow. I've read a lot of books but none that combined the three things that have driven my life. I'm one of those Quakers who sees the essence of life as bigger than the Bible, a scientist who has been invited to help with resource management and climate change issues in about 30 countries on 5 continents and a University professor who is witnessing more faith in younger generations than those of us who helped create current challenges to life on this planet. This mind-expanding book combines all three, sometimes unnervingly but definitely in a worth-reading way."
—Fred Bunnell, Professor Emeritus, University of British Columbia

"Sasso's book is a biblically inspired and scientifically literate presentation of the gospel of justice, creation care, and New Creation. Its message is the same as Pope Francis' *Laudato Si*, only in a youthful, more popular, and more apocalyptic Protestant voice. Recommended for both Christians and secularists alike."
—Herman Daly, Emeritus Professor, School of Public Policy, University of Maryland

"I've been searching for a book about climate change that is invitational, knowledgeable, theologically informed, engaging, and convincing. And now I've found one. Read this book and let it do its magic on you. It worked on me!"
—Tim Dickau, Chair of Justice and Mercy Network, Canadian Baptists of Western Canada

"Eduardo Sasso has given us his heart in this work. And, it is the heart of folks like him that can change the world we live in. It is my hope that Christians, especially in North America, will take his words to heart and live into the gospel of Jesus which demands our involvement in the healing of Mother Earth."
—Teresa Diewert, Community Organizer, Streams of Justice

"Although I have only met Eduardo a few times I have always been struck by his concern to integrate the varied callings, gifts and wide experiences that he has been given. His expertise as an engineer, his love of creation, and his theological training all come together to give us this personal and passionate book which I hope will find the readers it deserves, and inspire them equally to embrace a gospel for all creation, and for the whole of life."
—Peter Harris, President & Co-founder, A Rocha International

"Eduardo Sasso is a man of many talents, with a keen mind and passionate commitments. This is a highly original and creative synthesis of three topics not often discussed in the same breadth: sexuality, climate change, and Christianity. Sasso brings together a fascinating amalgam of facts, views and stories to propose a constructive way forward out of our present impasse. *A Climate of Desire* is a multi-disciplinary study written in a widely accessible way."
—Richard Higginson, Director Faith in Business, Ridley Hall, Cambridge

"Written in evocative and highly readable style, *A Climate of Desire* is an informative and up to date source regarding the pressing ecological issues of our times. Eduardo Sasso is a captivating storyteller who creatively addresses the challenge of ceasing to live carelessly 'chopping down nature's trunk, and instead live gratefully off the fruit of her branches.' Enjoy the read. Even better, join the cause."
—Pierre Lebel, Director, North America Cities Leadership Circle, YWAM

"Usually engineers convince with arguments, but this engineer is a powerful poet capable of making sense of what is going on in the world today and walking us through it. Eduardo's theology is sound, his research on climate change is factual. Please be advised as you pick up his book: once you start reading, you won't be able to stop."
—Norman Lévesque, Director, Green Churches Network

"Even for an avowed agnostic like myself, *A Climate of Desire* makes a compelling case we must draw upon all parts of the human experience—our minds, souls, and bodies—to meaningfully address the climate crisis."
—Asher Miller, Executive Director, Post-Carbon Institute

"*A Climate of Desire* is a passionate call to attentiveness and to change. Read it only if you are willing to experience doubt and be unsettled! But read it, too, if you want help in imagining a different way of living in this world."
—Iain Provan, Author of *Convenient Myths* and *Seriously Dangerous Religion*

A Climate of Desire

A Climate of Desire
Reconsidering Sex, Christianity,
and How We Respond to Climate Change

Eduardo Sasso

WIPF & STOCK · Eugene, Oregon

A CLIMATE OF DESIRE
Reconsidering Sex, Christianity, and How We Respond to Climate Change

Copyright © 2018 Eduardo Sasso. All rights reserved. Except for brief quotations in critical publications or reviews, no part of this book may be reproduced in any manner without prior written permission from the publisher. Write: Permissions, Wipf and Stock Publishers, 199 W. 8th Ave., Suite 3, Eugene, OR 97401.

Wipf & Stock
An Imprint of Wipf and Stock Publishers
199 W. 8th Ave., Suite 3
Eugene, OR 97401

www.wipfandstock.com

PAPERBACK ISBN: 978-1-5326-5551-7
HARDCOVER ISBN: 978-1-5326-5552-4
EBOOK ISBN: 978-1-5326-5553-1

Manufactured in the U.S.A.   08/22/18

At the author's request, all royalties from sales of this book will be given in support of the work of two ecological organizations: *Unist'ot'en Camp* in Canada and *FundeCooperación* in Costa Rica. For more information on these initiatives, visit: www.unistoten.camp and www.fundecooperacion.org. The author is not associated with these organisations in any form.

Scripture quotations are taken from the Holy Bible, New International Version®, NIV®. Copyright © 1973, 1978, 1984 by Biblica, Inc.™ Used by permission of Zondervan. All rights reserved worldwide.

www.climateofdesire.com

For all the earthkeepers and climate caretakers in North America—
especially those inhabiting the territories of the Salish Sea
in the Fraser River watershed.

"I wish it need not have happened in my time," said Frodo. "So do I," Gandalf replied, "and so do all who live to see such times. But that is not for them to decide. All we have to decide is what to do with the time that is given us."

—J. R. R. Tolkien, *The Fellowship of the Ring*

Contents

Acknowledgments | ix

Opening Words | 1

1 Sex & the Cities | 12
    A Tale of Five Voices

2 "Just Gimme the Facts" | 27
    Ecological Vital Signs in a Snapshot

3 Babel, Babylonia, & California | 46
    An Ancient Story, Remixed

4 Climate Change & the Good News | 67
    A Glimpse into the Ecological Vision of the Scriptures

5 Giving to Caesar What is God's? | 82
    Exploring Faithful Political Engagements

6 When the Climate Changed | 103
    Hope for Today

7 Becoming Earthkeepers | 119
    A Call for All of Us

Afterword | 137

Annex: An Ongoing Experiment | 139
    The Beginnings of a Concrete Response

Endnotes | 149

About the Author | 167

Acknowledgments

THIS BOOK IS A collective effort. More a journalistic essay than anything else, the pages below seek to honor and bear witness to the many earthkeeping practitioners who are leading the way.

I'm quite humbled by the gracious contributions of all who provided financial support for the printing and production of the book: King-mong Chang, Guido Goicoechea, Eugenia Goicoechea, Norma Goicoechea, Pierre Lebel, Stan Olson, Maggie Knight, Daniela Sasso, Raquel Sasso, and the rest of contributors who decided to give anonymously. My gratitude also to Anna-Liisa Aunio, Fred Bunnell, Shakina Khan, Kurtis Peters, Ahna Phillips, Hannali Popoca, Richard Renshaw, Peter Reynolds, and Mark van Bommel for their helpful feedback or proofreading.

Credits as well go to my intellectual mentors, many of whom I don't know in person but only in writing. To Richard Heinberg and Bill Rees for informing my understanding of urban thermodynamics, to Gary Hewitt and Paul Williams for helping me understand the religion of consumerism, and to Jeremy Begbie, Walter Brueggemann, Ellen Davis, Gustavo Gutiérrez, Jürgen Moltmann, Iain Provan, and N. T. Wright, for illuminating the artistic and liberational relevance of the biblical writings and of Jesus of Nazareth for the twenty-first century.

On a personal level, I'd like to express my sincere gratitude to the testimony of Mary-Ruth and Loren Wilkinson, from whom I've borrowed the term "earthkeeping," and in whom I've seen it practiced. Thanks as well to my earthkeeping friends in Vancouver, for speaking up and stepping out. And also a many thanks to Teresa and Dave Diewert, Mar Barquero and Migo Álvarez, and Erin and Kurtis Peters, all of whom have been examples and sources of encouragement to me, each in their own peculiar way.

Acknowledgments

Last but not least, I would also like to extend a sincere and special 'thank you' to Megan Fraser, whose graduation project on levity and gravity saved me from myself, and to my colleague and friend Jason Wood, who embodies and exemplifies pretty much everything this book stands for.

Amigas y amigos, mil gracias a cada uno de ustedes . . .

Opening Words
Maple Syrup, a Beach in Costa Rica, and the Age of Fossil Fuels

> "Not everything that counts can be counted,
> and not everything that can be counted counts."
> —Albert Einstein

> "See, I am of small account; what shall I answer you?"
> —The Book of Job (40:4)

IT TOOK JUST OVER forty-eight hours. After checking my luggage at Vancouver International Airport, in a little less than two days I found myself with some friends at Playa La Prieta, a beach in Costa Rica's westernmost province. The place gives one a true glimpse of paradise. It was late December of 2009—a season of rain and snow up north, but one of just about the perfect weather on the coasts of Central America. Difficult to find a better way to end the year after almost sixteen months of living abroad. I had been breaking my head as a grad student aiming to blend business with ethical and theological studies, so I truly welcomed the holiday. In fact, being at La Prieta felt like rediscovering my native land all over again, feet-sunk in the golden sands that shifted mildly every time the ocean waters splashed in. The sight and sound of a few distant gulls added to the much-needed time of rest.

After the quick dinner that followed our four-and-a-half-hour drive from Costa Rica's capital city, San José, my friends and I happened to round off the first night's meal with some vanilla ice cream, which has never been my favorite. However, my mind lit up when I remembered the bottle of 100 percent maple syrup I had brought with me. Real maple syrup is something my Canadian friends take for granted, but it remains a treat for us Costa Ricans. So, thankfully the dessert became anything but dull after dazzling it with my supply of golden taste.

"Do you realize the irony?" a friend asked, eventually.

"Of what?" I replied.

"We're savoring our ice cream topped with this fancy delicacy you brought all the way from Vancouver . . . but that's possible pretty much only because of fossil fuels."

"Aren't we blessed!" someone else remarked.

A while ago, one of us had mentioned the likelihood of rivers flooding and sea levels rising given recent changes in the earth's climate. But there we were on a pristine beach, overwhelming our sight with a radiant sunset and relishing our taste buds with genuine maple sweetness. The goodness was so real; the likelihood of rising seas felt so distant. For all we knew, the age of fossil fuels was on our side. It took Christopher Columbus over sixty days to bring scissors and gadgets across the Atlantic, but there we were delighted with a charming bottle that flew halfway down the world's longest continent in just around forty-eight hours.

We were blessed indeed, I thought to myself, even if the magic syrup left me with an unknown aftertaste that never fully went away.

Wasn't Christianity Out of Fashion?

This book is an invitation to reconsider sexuality, Christianity, and what we do and don't do about climate change. At its core, it's also a book about desire: about what we long for and about the consequences of our longings.

Speaking of these four issues together is surely an odd combination. It's also likely improper or uncomfortable to secularists, agnostics, and Christians alike. Odd, improper, or uncomfortable because these are realities we all feel strongly about and often disagree on, in one way or another. Why bring them together?

The oddity deserves an explanation. So allow me to begin with a quick remark about faith to make better sense of why sex, Christianity, and climate change are more intermingled than we often imagine.

Opening Words

As may have been the case for many during their university years, by reading nineteenth-century intellectuals like Ludwig Feuerbach, Friedrich Nietzsche, and Sigmund Freud, I was taught to take for granted (almost religiously) that God was an illusion invented by our primitive ancestors. These thinkers argued that, way back in time, humans created a heavenly figure, or figures, in their image and then painted such images into the blank canvas of the cosmos. The mythical images served numerous purposes: to try to explain the otherwise mysterious forces of nature; to comfort the infantile desire to have a strong father figure; to control, appease, or terrify others; and so on. However, these intellectuals claimed that with the advent of modern science, and with the deeper study of the mind, the West arrived at a point where it could finally dispel old religious fables. I was thus taught to think that Christianity, Islam, Hinduism, and Roman or Egyptian polytheism were all a *myth* and an *illusion*—realities that solely exist in people's imaginations but not outside of them.[1]

Knowingly or unknowingly, such beliefs have more recently stirred artists and scientists alike, all the way from John Lennon to Richard Dawkins. The first proclaimed there were no heavens but only skies above us; the second has continued to promote a neo-atheist gospel, fervently arguing that God is a delusion and faith is a virus. In turn, would-be journalists like Dan Brown have written popular books like the *Da Vinci Code* to persuade millions that the New Testament is little less than a power-grab and a farce. And even more thoughtful historians, such as Yuval Noah Harari in his best-selling *Sapiens: A Brief History of Humankind*, have presented it as a matter of fact that God, the UN Declaration of Human Rights, or human equality, are ultimately nothing but inventions of the human mind.

I came to acknowledge there was something true about such critiques of religion. King Hammurabi from the old Babylonian Empire affirmed that the gods had charged him with dividing society into free men, commoners, and slaves. Armed to the teeth, the Romans claimed and made others believe that the gods were on their side, allowing them to subjugate the imagination and collect taxes from over 100 million subjects. Many self-professed Christians have sometimes done likewise, often to the point of holding the Bible in one hand and a sword, a rifle, or a MasterCard in the other. All that to agree that our collective and personal histories do prove that it's rather convenient to believe in a God made in our image, often obedient to our selfish whims or submissive to the wellbeing of our tribe. Religion can indeed function as the opium of the people.

But are things as simple as that? Given that we're only able to draw imperfect circles on a wrinkled scrap of paper does not immediately imply that there is no such thing as a real three-dimensional sphere, to paraphrase Irish writer C. S. Lewis. A five-year-old tapping clumsily at a piano says nothing about whether Beethoven did or didn't actually write the Ninth Symphony. The fact that many, or most, of our understandings of God are either partial, selfish, distorted, or false does not automatically imply that God (whoever God may be) *must* be a mere illusion. It might well be that the cosmos has an incredibly patient and yet ultimately untamable living source trying to meet us where we are at by appealing to our earthly metaphors, as to speak to us in familiar language about such a far-superior presence that transcends us. And it might be the case that such a vast presence is there whether we believe in it or not. A total eclipse or a perfect storm does not mean that the sun has in fact disappeared.

I write under such convictions. First, I am convinced I have somehow experienced such presence in my own life—very much like someone who experiences a musical tune he cannot see or touch. And secondly, I'm convinced that instead of rejecting God from the get-go, true and honest openness of mind calls us to at least acknowledge the *possibility* that there is a God who we might not be fully able to grasp, like, or understand.

Regardless of our religious or philosophical inclinations however, anthropologists and historians continue doing their best efforts to remind us that our past largely determines our future. For better or worse, Christianity has significantly shaped cultures in the West for the past 1,700 years or so, and it continues to do so today when over 33 percent of the world's population affirm some sort of allegiance to it. But by ignoring or denying that past, we can fall into a sort of collective Alzheimer's: our societies can become confused about who they are, or simply inclined to repeat old mistakes. Forgetting our history is usually a guaranteed recipe for social dysfunction. What is more, studying the earliest history of Judaism and Christianity may in fact surprise us, revealing a rather different God from the one we've been taught to believe in—or disbelieve in.

Likewise, the rejection of faith and traditional values that marks our increasingly urban societies runs the risk of throwing away the baby with the bathwater. Such reaction leaves us either muddled, helpless, or empty-handed as we walk into a post-secular age, when we're craving deep meaning and genuine spirituality. Despite most of our efforts to the contrary, Bono's song continues to linger in the back, as we recognize that we still haven't quite found what we're looking for. Our desire for 'something more' is alive as ever.

In turn, whether or not we identify ourselves as persons of faith, and whether or not we believe that we humans are the leading cause of climate change, these are realities that are not likely going to go away. So even if one dislikes science, Christianity, or religions in general, they require attention. One might not entirely agree with people of faith (or with people of no faith), but at the end of the day, our increasingly agitated world continually calls us to collaborate across boundaries—especially when it comes to addressing major challenges such as those related to the ecological breakdown our world is experiencing.

An Engineer's Melting Faith

That call to collaborate is one my reasons for writing. As an engineer who at different times of his life has been both a secularist and an admirer of the intriguing person of Jesus of Nazareth, here I'm bringing secularism and Christianity into dialogue. I try to do so somewhat seeking to honor the words of Albert Einstein, who recognized that "science without religion is lame, [and] religion without science is blind."[2] In fact, this book is a response to different call outs to religious communities, such as the one posed as early as 1992 when the Union of Concerned Scientists urged for the integration of faith and science around new ways of caring for ourselves and for the earth.[3]

Why is this integration needed?

Allow me to be self-referential again in trying to unfold the importance of blending faith and science. Being the son of a civil engineer and grandson of two civil engineers, I went to university to equip myself as an industrial engineer. Having been born and raised to be rather box-minded, I received further training to solve problems: obtain information, know what set of procedures to apply, throw in the data, and done: QED. Crystal clear answers. For me, life was very much an ordered procedure where one could control all the variables and eventually arrive at one concrete, stable solution.

After working for three years in the textile industry in Costa Rica however, around 2005 I began to become mindful of the potential of my discipline both to do good, but also to cause harm. Enterprises like Patagonia or The Body Shop inspired me with their progressive vision and compassionate leadership styles, but I soon realized that companies like those were comparatively few. Eventually, too, documentaries such as *Food Inc.* and *The Corporation* exposed me to the hidden side of business. Opening the newspapers was like reading *The Hunger Games*. My own awareness of the polluting effects of textiles brought me to the recognition that my discipline was in need of serious repair. I particularly remember one time I visited the cotton fields in South Carolina during a business trip sponsored by the National Cotton Council of America; it felt like going to a war memorial cemetery, save for the fact that the victims were cotton plants plastered with synthetic pesticides. There was literally no trace of life except for countless square miles of white puffs.

As someone who had always been struck by the intricacy and splendor of our planet, and as someone who had come to recognize the world as the living artwork of a far superior living being, realities like these meant that my tidy formulas and engineering diagrams stopped making full sense.

Fast forward to the Fall of 2012, a time when two friends and I had the chance to drive across the Canadian Rockies. On our way back from Canmore, Alberta, we stopped at the outskirts of the Athabasca glacier in Jasper National Park. At first, I was gladly surprised when I looked ahead and saw a sign on the ground that had my year of birth on it. But as we moved closer, a friend read the small print: "The glacier was here in 1982." My surprise turned into shock as I heard a stream of melted ice water, soon realizing that the glacier had receded over 600 feet in my own lifespan and around 1.25 miles since its discovery in 1844.

"Was that because of climate change?" "Are 600 feet that bad?" As an engineer, I wondered; as a person of faith, I felt small and perplexed. And then I remembered maple syrup and my conversation at the beach.

The Economy, Christianity, and Climate Change

Eventually, situations like these made me wonder about the relationship between climate change and the cultural climate cheered for by the current economic system that now shapes most of the world. While making goods like maple syrup instantly available across a continent, I realized our fossil-fueled economies often have an untold story, which we'll explore in the pages below. But regardless of whether climate change was real or not, as someone who had some awareness of the biblical tradition, my industrial engineering consciousness was agitated further soon after when I came across American Christian writer Wendell Berry, through his famous essay "Christianity and the Survival of Creation."

Some modern Christians, Berry wrote, are complicit in the ecological villainy that comes by either believing that Christianity is about an inward spiritual journey that leaves the world unchanged, or about sending weak souls to heaven. Provocative as he is, Berry went on saying that the complicity between Christianity and our predominant economic system is in fact largely responsible for what he called "the military-industrial conspiracy to murder creation."[4]

That rang a bell or two. To my knowledge, both ancient and modern expressions of Christianity had spoken lots about God but rarely about the land or the earth. In fact, some Christians have often (mis)read isolated one-liners of the Bible to find quick justification for the domination of animals and of nature. Some historians have hinted that, as early as the Middle Ages, Christians have relied on the biblical call to exercise dominion as a rather convenient excuse for being at ease with inflicting violence on the land and its creatures, whether slicing and turning the soil with heavy ploughs, overharvesting, chopping trees, or butchering pigs.[5]

I knew, as well, that several expressions of Christianity have often scorned or denigrated sex and that they have been complicit with a good deal of the sidelining of women. Influenced by Greek Neoplatonist philosophies that elevated the soul above the body, since early on in the church's history, numerous Christians came to see the body as evil, pleasure as sinful, and sex as shameful. And despite the New Testament's subversive affirmations that the distinctions of worth between men and women were set aside by Jesus, Christianity eventually assimilated the patriarchy pervasive in the culture of the Roman Empire. Men alone were to rule the show.

On the other hand, my university years made clear that much contemporary humanism tended to sanctify sex. Over a century ago, Freud called for unleashing suppressed sexual desires. Today, many have magnified the plea and followed suit, be it merchandisers or countless pop stars, be it professional psychologists or Lady Gaga. Likewise, even if the Renaissance came to focus its spotlight on humanity largely at the expense of everything else, eventually secular humanism began to recognize the importance of the land and the earth. And yet (and perhaps rather religiously) secularism continues to show little signs of allowing for the existence of a cosmic spirit, an ultimate consciousness, a living source, a unifying presence—let alone of God.

A Climate of Desire

Furthermore (and here's the rub), Christianity and secularism hardly speak of sex as a political relationship, let alone as an ecological one. In our me-centered world, we understand sex as an interpersonal act restricted to bodily intercourse. Rarely, however, do we speak of sexuality as the way cities and human communities relate to one another or to the earth itself.

But what if sex has more political and ecological ramifications than we may have come to believe, as the glowing speech of Jeremiah and other ancient seers will provoke us into considering come chapter 1?

Maybe our desires and our sexuality extend to our social and ecological interactions as well, with the latter serving as a metaphor for relationships of affection and reciprocity that are grounded in grateful receiving and self-giving, and not in endless extraction or in the mere taking for selfish pleasure. Perhaps sexuality is the deepest, all-encompassing language that most fully captures the right relationship between the spiritual and the physical, between heaven and earth, between God, humankind, and the entire cosmos.

∼

In the midst of tensions like these, some years ago I was gripped by a question: What if popular culture, secular humanism, and Western Christianity are one-sided and, ultimately, incomplete? What if the earth calls us to orient our deepest desires towards a transcendent, originating living source—in this case, the living source whom the descendants of the ancient Hebrews recognized as *Yahweh*? And what if Yahweh's *ruach* is the ultimate breath of life, uniting, healing, and inspiring humans across race and language to be attentive to the earth and to each other?

These pages are an exploration along these lines, aiming to reconcile the religious and secular worlds. They are also an experiment seeking to transcend the narrow, this-leads-to-that linear logic that is so prevalent in our Western world. I am, after all, a recovering engineer grasped by the challenge of British writer G. K. Chesterton, who recognized that "the poet only asks to get his head into the heavens. It is the logician who seeks to get the heavens into his head. And it is his head that splits."[6] That is to say that I write convinced that for us to experience and to tune in with the mystery of the divine passion—the pathos—that runs through the fabric of the cosmos, we're in need of other ways of engaging with all that surrounds us. While ideas have legs and arguments have feet, stories have hearts, poems have wings, and songs have dancing shoes.

Opening Words

Risking being misunderstood, however, I welcome the patience of both secularists and Christians alike. I've taken the risk of trying to lay down what I hope is a more holistic—though unfamiliar and very likely uncomfortable—pathway that may enable us to better respond to the challenges of today. Even though I write both as a theologian and an engineer, I'm also aware that in our modern (and now postmodern age) religion and so-called secular science have been held apart from each other. Thus, I write convinced of the urgency to reconcile these two perspectives—and in some cases even transcend them when they fail to serve us. In fact, in unfolding the relationship between sex, Christianity, and climate change, the essay below is a blend of insights from other theorists and practitioners as well: economists, storytellers, enlightened CEOs, biblical scholars, artists, pastors and ministers, business consultants, community organizers, historians, journalists, social scientists, ecologists. The scale of the challenges of our century calls for multiple instruments to come together around a new song of cosmic healing, justice, and peace. So this is my best attempt at interpreting what the musical score might read like, knowing that it will be up to each of us to join others in making the music come alive, however imperfectly.

What to Expect from the Chapters Ahead?

The pages below move back and forth between realism and imagination, practice and possibility, analysis and anecdote. At times, you'll hear an engineer writing, discussing climate stats and connecting the various dots; at other times, you'll encounter a storyteller in the making, trying to remember the past as a way to understand our present moment. Some chapters are more research focused, others read more like a piece of creative writing. However, rest assured that I'm ultimately no specialist in the fields I touch upon. What you'll meet below is a popularizer of sorts, doing his best to blend and translate for all of us the insights of different experts and practitioners.

In the Annex I'll share some of my own involvement in the foundation of Earthkeepers—a recent network of people of faith in the unceded Coast Salish Territories, Vancouver, Canada, seeking to respond creatively to the challenge of climate change. There I lay out some of the grounds for what we together discerned could be a promising response to such challenge that was rooted in the tradition of public love, central to some expressions of

Christianity. In turn, chapter 4 unpacks more closely some of the ecological ramifications of the Hebrew and Christian Scriptures, exploring what an ancient poet might have meant when he affirmed that "the earth belongs to Yahweh and everything in it" (Ps 24:1).

In that spirit, chapter 1 begins with an explicitly unscientific introduction to our tour guides—two protagonists and three supporting characters that will pop up here and there throughout the book. They will provide us with the provocative metaphors that will guide our imaginations. Following, chapter 2 gets more analytical, moving on to explore some of the vital ecological signs of our living home, planet earth. Chapter 3 unfolds some parallels between the ancient story of Babel and our contemporary story of globalization and consumerism, aiming to unveil some of the causes of climate change. (Readers are encouraged to press on through these three opening chapters, which may be more unnerving than most of us may be used to.)

In response, the relationship between climate change and the often-ignored ecological vision of the biblical writers is sketched in chapter 4. There we will revisit the sense of cosmic sexuality—of what I believe could be called sacred intercourse—as portrayed in the Apocalypse, the last book of the Bible. Then chapters 5 and chapter 6 outline some concrete responses to the climate challenge, as well as some historical reasons for hope. There I draw inspiration from previous movements of social change spearheaded by faith communities—namely the nineteenth-century abolitionist and twentieth-century civil rights movements. The book ends with chapter 7, making a call to stretch the horizons of our hopes while embracing a posture of humility and earthkeeping.

It's perhaps worth underlining that this book is written for anyone curious (or at least open) to these topics. Whether secular, agnostic, Christian, or otherwise, I don't expect all readers to share my faith or political convictions, nor do I intend to convert others to them. In fact, I uphold and admire the contributions and the example of many people who don't profess alliance to any particular religion, let alone to Christianity.

My intention here is simply to recover some forgotten voices that I believe continue to be either silenced, misunderstood, or forgotten, but that can nevertheless shed provocative light and provide inspiration as we face

OPENING WORDS

our shared challenges ahead. So even if I write as someone rooted in the Judeo-Christian tradition (because of which, I painfully recognize, much blood and hatred continues to be spilled), the pages ahead are open to anyone who resonates with the ancient vision of a cosmic throne that belongs to a Lamb who reigns in meekness and peace.

1

Sex & the Cities
A Tale of Five Voices

"There are more idols than realities in the world..."
—Friedrich Nietzsche, German philosopher

"Sir, my concern is not whether God is on our side;
my greatest concern is to be on God's side, for God is always right."
—Abraham Lincoln, former Republican President

CHRISTIANITY, SEX, AND CLIMATE Change. It's challenging to identify more provocative or controversial issues than these today. For one, the eccentricity grows when the three are brought together, as this book attempts to do. But the oddity grows exponentially when one suggests that the three are woven together, and in more ways than one. So, to state it again, I'm well aware of opening Pandora's Box by making this claim because there is indeed a bittersweet history behind sex, Christianity, and the burning of fossil fuels, which, according to some, is leading to climate change.

Since the days of the Romans, Christianity has been a source of encouragement, bringing hope and renewal to millions, sometimes to the point of inspiring the creation of public hospitals, the universal establishment of schools, and the launch of abolitionist and civil rights movements. In turn, the Christian faith and its holy book have also been distorted and abused. Not long ago, many Nazis used the Bible to validate their persecution of the Jews and their support of Hitler. Europeans used it to baptize their swords and rifles as divine instruments to colonize Africa, Asia, and the Americas in the name of God.

Secondly, global warming. Recent changes in the earth's atmosphere and oceans are hurting millions today through droughts, forest fires, and floods. Still, the possibility that it's us humans who mostly cause climate change continues to be doubted, downplayed, or outright dismissed. Some, for instance, have confidently portrayed it as a convenient fraud, supposedly invented to justify the mass production of solar panels. The popular claim is that climate change is very much a lie invented by China to debunk the economic supremacy of the coal, oil, and gas industries in the United States and elsewhere.

Thirdly, sex. Well . . . sex is sex, and for all of our bitter or sweet experiences with it, and without its sublime spark, hardly any of us would be here.

How can these three powerful, and potentially dangerous, forces be related, if at all?

We could begin to answer the question by taking the much-traveled path of exploring stats and figures related to the state of our living planet (some of which we'll do in chapter 2). However, to start there would be putting the cart before the horse, and it would be to restate what is becoming more and more obvious.

Instead, we'll begin in a rather odd place by opening our imagination to the visions of two eclectic seers of old: Jeremiah and John. Outlandish and bizarre today as they were back then, they will be our main guides leading us through the rest of our journey.

Two Daring Visionaries

The Big Mouth of Jeremiah

In the seventh and sixth centuries BCE, a young Middle Eastern prophet by the name of Jeremiah felt called to speak to his contemporaries, the Jewish people of old—a group of people who saw themselves as specially chosen by Yahweh.

Very long story short, the Jews came to confess Yahweh as the one true God and as the ultimate living source and liberator of the cosmos. Some of them also saw themselves as uniquely elected for the sake of bringing light and salvation to the nations. The descendants of the ancient Hebrews came to believe that Yahweh had set his eye on them to bless them and to bless the entire world through their offspring.

As it happened, their election was depicted rather candidly. They came to see their relationship with the Most High as nothing short of a covenant of marriage. Jeremiah recalled the times when the descendants of the Hebrews were like the bride of Yahweh. The young visionary spoke of the days when Moses led the Israelites in a rescue operation from servitude under Egyptian rule to then settle on a strip on land in rugged Palestine. After being landless slaves without identity in the land of empire, marriage with Yahweh meant a new life in a land of promise.

As centuries passed, however, most of them forgot their origins and their collective history, and thus came to take themselves—and their God—for granted.

Thus came forth Jeremiah. Wearing his heart on his sleeve, the young visionary felt himself called to remind the ancient Jews how their special election came with special stipulations to be observed on a special land with reference to their special God, Yahweh. The former rescue operation out of Egypt was never intended to be a blank check; the people of Israel were simply not meant to do as they pleased. The promised blessing was always conditional on living righteously in a promised land that was not theirs.

Tragically, the chosen people hardly lived up to the standards. In fact, most came to see the gift of land as cheap grace and turned Yahweh into a sort of domesticated puppet within their reach and control. Having been freed from the economic slavery of eight-day work weeks, the people of the covenant went on to labor on the days of rest; a few of them accumulated wealth and built their buildings and palaces on the backs of the poor; most of them treated the land as a discounted commodity. They ignored the stipulations. They broke their pledge with Yahweh.[7]

Being the spokesperson for the land that he was, Jeremiah saw it coming. From Moses, he had learned that the land would reject its invaders, and thus Jeremiah claimed to have fresh words from above put into his mouth—words that were perhaps too hard to swallow back then, just as they are now.

The descendants of the Hebrews, Jeremiah said, had become like a "wild donkey" craving after the sustenance of foreign gods. They had bowed down before idols engraved in wood and stone, seeking riches but neglecting justice, oppressing the foreigner, and shedding blood of the innocent poor (Jer 7:5–6). Both religious and national leaders had closed ears and stubborn hearts; they were false preachers of peace when there was no peace (6:10, 14); they were confident that everything was going to be O.K. The descendants of the Hebrews, Jeremiah continued shamelessly, became good at "lying down as a prostitute under every spreading tree" (2:20–25). They became "skilled in pursuing love" (2:33) like a "prostitute with many lovers" (3:1). The prophet had reasons for his scandalous declarations: Even as they turned their backs on Yahweh, the people of the covenant opened their legs to the wealthy ways of the gods of empire. They engaged in what could be called an ancient version of an international hook-up: money, power, and political sex.

Such was not a popular message to spread, giving Jeremiah reasons to resist it. And he did at first. But the living Spirit stirred within his guts with words too large be contained, leaving the young prophet without an option. The unholy love affair with false deities demanded the lifting of the blankets to expose Jerusalem's adultery. Faithfulness to Yahweh meant a life of justice; desiring other gods meant the opposite. Burdened with grief, Jeremiah knew the special gift would be lost; he knew that the land would "vomit out" those who defiled it.[8]

Following the candid verbal show, some likely labeled him a sensationalist, a radical, a weirdo. And yet what is true is that kings, pastors, and court prophets persecuted him, calling him a national traitor for speaking truth to Jerusalem. They beat him, imprisoned him, put him in stocks.

But persecution, anguish, and dismay could not keep Jeremiah's big mouth from being shut—even if Jerusalem would not listen. The people's vision was short-sighted, keeping them from noticing the hook within the bait. Their lovers enticed them if only to obtain their booty:

> Why dress yourself in scarlet
>   and put on jewels of gold?
> Why highlight your eyes with makeup?
>   You adorn yourself in vain.
> Your lovers despise you;
>   they want to kill you. (4:30)

The reason for all the incandescent words? For one, the descendants of the Hebrews had forsaken Yahweh, the living source. Led by greedy kings, they also drunk from the artificially flavored pop drinks sold by their surrounding empires. They worshipped false gods and embraced heartless economic practices, which eventually turned Yahweh's promised strip of land at the East end of the Mediterranean into something detestable. The nation's leaders were at ease with the blood of "the innocent poor" (2:34), they had eyes for "dishonest gain" (22:17), and they "polluted" and "defiled the land with their vile whoredom" (2:7; 3:2, ESV).

In his book, *The Land*, scholar of the ancient Hebrews Walter Brueggemann notices how the chosen people had "ceased to trust the land-giver and engaged in alternative ways of securing their own existence." Brueggemann continues: "[their] disobedience and carelessness not only offends Yahweh . . . it also affects the land. The land has its own life and its own meaning . . . It is the land that is finally abused. Harlotry is the way to lose the land."[9]

Jeremiah's diagnosis was razor sharp: ignoring Yahweh led to abusive behavior toward the neighbor and to careless stewardship of the land—the two sides of the same old imperial story. And the love affair horrified the prophet, leading him to insist on raising one red flag after another. Even if the rich amassed the land and controlled it as disposable merchandise, Jeremiah nevertheless called his people to treat it as a sacred gift that came with small print attached.

Still, the indulgence continued despite the countless admonitions. Called to be a city of peace, Jerusalem did its best to reflect the contrary, all the way to turning their sacred temple into what the prophet revealed to be "a den of thieves" (7:11, NLT).

But both seers and prophets knew that Yahweh would not be mocked. The sacred writings were crystal clear: obedience to the covenant led to blessing; sleeping around with false gods, did not. And so, after many centuries and countless warnings, on went the city's people, and thus they were "vomited" into exile. Having once wandered in circles in the desert, the land-given people became, once again, land*less*—but now captive in the ancient empire of Babylon. Their journey into exile was preceded by adulterous acts of false worship and misguided political sex.

The unimagined possibility of the greatest-ever national tragedy did in fact make the headlines: the land was lost. And in the midst of it all, Jeremiah embraced the cry of Yahweh:

> Let my eyes overflow with tears
>   night and day without ceasing. (14:17)

Sex & the Cities
John's Apocalyptic Romance

Jeremiah was not alone. In the first century AD, a seer by the name of John followed suit. But unlike the sixth-century prophet who spoke for the most part to the descendants of the Hebrews, John's revelations unveiled the multicolored mischief of all the kings of the earth. For one, John was a curtain-lifter and a whistle-blower; a man with eagle-eyed vision able to cut through the neon lights and the merchandising smoke screens of Rome's empire and of all others following.

As of late, however, John's voice has been greatly silenced. Some best-selling twentieth-century authors have turned him into a crystalball guru of sorts, supposedly foretelling the exact sequence of supposedly disastrous events in a distant future.[10] (All their predictions have failed so far, even if such books have made their authors lots of money!) But, surprise, surprise: John was an encourager, not an end-times doomer. Inspired by the prophets of old, John reinterpreted the Hebrew Scriptures around a crucified king, whom he came to see as the "ruler of all the kings of the earth" (Rev 1:5). Thus he wrote the climax of prophecy: a set of visions to challenge and inspire communities of faith to "endure patiently" and "hold fast in their witness to the Messiah" (3:10; 14:12).[11] John also wrote to assure them that the throne of the cosmos belonged to a humble Lamb who had been slain, and not to puffed-up, beastly powers that claimed to own it (chs. 4–5; 18). Hope, not doom, was the order of the day.

Just as they do today, claims to the throne of the universe meant jealousy and conflict. Millions considered the entire commercial and military apparatus of Roman culture as the cutting edge of civilization, with all their roads, aqueducts, theatres, temples, commercial inroads, and the largest and most intimidating army of the ancient world. Rome's might and propaganda intimidated and enticed all nations around the Mediterranean. The great city lured countless kings from everywhere to follow her and to hand over their power and authority. Some kneeled submissively because of fear; others surrendered wholeheartedly despite the compromise. In seeing Rome's grandeur, even John marveled greatly (17:6, 13).

In the blasting vision given to him, however, the ultimate rule did not belong to the nations and their leaders, nor to Rome and its emperors. It was the slain Lamb who was the King of kings. John was to make people aware that Rome was the beast. The great city was a first-century remix of the old empires of Egypt and Babylon, but this time flying high on wild mushrooms and fancy steroids. Rome became the transnational reembodiment of the imperial powers that had always been at odds with Yahweh, the ultimate living source. Rome was "the great prostitute" with whom "the kings of the earth committed sexual immorality" (17:2, ESV).

As with Jeremiah, John could not portray the drama with dry, polite speech when he described the harlotry:

> For all the nations have drunk of the wine of the wrath of her fornication, and the kings of the earth have committed fornication with her, and the merchants of the earth have grown rich from the power of her luxury. (18:3, NRSV)

With one hand, John unveiled nations and kings entangled in political dealings with the world's self-boasting superpower; with the other, he revealed the unquenched desires of merchants involved in transnational unfair trade agreements. The blacklist of luxury items was anything but short. Trading with Rome meant unending "cargoes of gold, silver, precious stones and pearls; fine linen, purple, silk and scarlet cloth; every sort of citron wood, and articles of every kind made of ivory, costly wood, bronze, iron and marble," on top of "human beings sold as slaves" (18:12). In short, adulterous acts of political and economic intercourse, said John. As with Jerusalem, so with Rome: money, power, and commercial sex. Lots of it. And with an unhappy hangover soon to follow, because "all this wealth has been laid waste" (18:17, ESV). After the glowing rave, the great city would be paid in full, John foreknew—a destiny confirmed to various degrees by several centuries of Medieval ruin, pestilence, and strife following the dissolution of the Roman empire in the fourth and fifth centuries AD.

After her multiple international hook-ups, Rome fell.

∼

Evidently, Jeremiah the prophet and John the seer were not trained in diplomacy or well-mannered modes of speech. Their sexual rhetoric was as charged up then as it is today. Still, what is perhaps more striking to our civilized sensitivities is that most of us see sex as an act between individuals. But these two ancient disturbers of the *status quo* urge us to reconsider. Like every other empire ever since, Rome was no more than "Babylon the great," "the mother of prostitutes and of the abominations of the earth" (17:5). For Jeremiah and John, sexuality was (and continues to be) also a matter of political, economic, and international intercourse. Enticed by all sorts of gods, cities also have sex.

What might any of this have to do with climate change? For one, in chapter 3 we'll see how the global economy is, in more ways than one, a re-appearance of "the beast," and that climate change is, in fact, a beastly "crisis of crises," to borrow the phrase from American ecologist David Orr.[12] Just as Rome's spiritual propaganda seduced entire nations, leading their elites to exploit the poor, live luxuriously, and "destroy the earth" (Rev 11:18), so is our multinational economic system drunk with similar vices.[13] Today's global party is run by fewer and fewer banks and corporate bullies, the tunes being played dance to the rhythm of consumerism, and the merchants of advertisement pass around countless silver trays with sparkling drinks offered for our mindless entertainment. Most guests have already passed out, though some continue to enjoy the feast. Regardless, the fireworks are almost over, and we're close to running out of cheap gunpowder.

As with Jeremiah and John, however, storytelling and poetry over-grow discourse and prose. Before further exploring the intermingling of sex, Christianity, and climate change, we come across a supporting set of characters: Boris Waterlove, the Charitable Doublelives, and the Floating Clowns.

Our Situation Today in Three Parables

Inspired by Jeremiah and John the seer, Boris Waterlove now takes center stage. The setting: a hot summer day at Boris's house; the context: a visit from his friends, Jillian and Jack.

Tired as they were, Jillian and Jack asked for a glass of crisp, cold water. To his shock and surprise, Boris found a jar of pink rat poison next to the faucet. Being the noble and righteous man that he was, he immediately removed the container from the kitchen counter. Boris wanted to make sure that not a single, minuscule drop of poison fell into their glass. Allowing that to happen would obviously not be a loving action. Like any decent person, he wanted to serve clean, fresh water to his friends.

A few days after, Mr. Waterlove found out that the public water system was being polluted. (Willingly or unwillingly, consciously or unconsciously, he did not know.) Still, he wanted to be confident that the municipal waters were crisp and pure: Jillian and Jack would appreciate fresh water, as would he.

Seeing himself as a devout Christian, Boris was conflicted. He knew that he was called to love his friends as he loved himself. But . . . should he go beyond removing the pink rat poison jar from the counter?

"Environmentalism is such a 'secular' thing," he had come to think. Regardless, the question lingered, "How can I have certainty that our municipal water is drinkable?"

Different possibilities stormed through his head. He could elect righteous officials to represent his interest in public office; he could participate in restoring the local watershed; he could even support a blue NGO, or perhaps he could even run for mayor (or for president). But . . . really? Giving his friends a glass of clear water sounded like a very holy thing to do, but the rest sounded . . . strange. Someone had taught him that Christianity was about loving people, not about politics or environmental issues.

To his own surprise, Waterlove had a conversion experience. Coming into awareness that the God whom he pledged allegiance to was also the living source of all things, Boris came to see the wider connections. He realized that in protecting the watershed, he was also protecting his own friends. Waterlove became convinced that "earthkeeping" was a way of loving thousands of people at the same time. Letting justice and clean waters flow down like a river was a way of walking the walk.

"Maybe it's true," he thought to himself, "that ecologists are in the business of love and uncommon sense."

∼

Jillian and Jack Doublelife, however, were not as excited. In fact, they were secretly upset with Boris's newfound enthusiasm.

Both Jillian and Jack happened to be children of majority shareholders in a transnational oil and gas company. The corporation sold what is popularly known as bitumen (crude oil mixed with dirt and sand) and liquefied natural gas (an exotic form of fossil fuel captured by penetrating and fracturing the intricate layers of the earth's crust).

To Boris's ignorance, the extraction of bitumen and the production of fracked gas (or, to use the politically polite term, "LNG") required voluminous amounts of energy and water: energy to heat the bitumen and cool the gas, water to crack layers of underground rock, water to pump the gas all the way up to the surface. In turn, building oil and gas pipelines also threatened hundreds of rivers, lakes, beaches, and oceans.

The Doublelives were vaguely aware of the ill effects that pipelines, bitumen, and fracking had on local waterways. Still, they managed to ease their consciences with a rather clever charitable trick. While making a profit by serving an energy-hungry economy with one hand, the Doublelives donated 10 percent of their income to water-cleansing and water-bottling charities with the other—a sort of philanthropic yin-yang. (The trick came in handy, now that their friend Waterlove had become a nut-head of sorts, preaching what they perceived to be a liberal green gospel.)

Boris felt rather helpless. Not that Jillian and Jack had bad intentions, though; not that they were greedy, or careless. In fact, they were nice, decent people. But they were born into a dominant system governed by what had become the standard business norm: the unspoken but ever-present imperative to sacrifice the Earth at the lofty altar of Mr. Profit.

The imperative was rather capricious. It required comfortably living off of chopping down nature's trunk instead of living gratefully off the fruit of her branches. The imperative meant extracting and consuming today at the expense of those coming tomorrow. It made everyone think that the world belongs to a few human beings. It called for the endless "penetration" of new markets (sexual pun unavoidably intended). The imperative made it imperative always to obey imperatives because almost everyone else was doing so.

The Doublelives were hard-pressed, indeed, by lots of imposed obligations. Hard-pressed because they didn't choose their last name, after all. It was forged and bestowed on them by a society that bowed down to glittering gods that have eyes that cannot see and ears that cannot hear. Outside of meeting Boris and his new set of green convictions, the Doublelives knew no better.

But, by sheer grace, the Floating Clowns did.

~

The Floating Clowns knew that watersheds were being penetrated and fracked up even as the atmosphere was changing. They knew about Boris's new enthusiasm and helplessness, and about the Doublelives' immersion in the imperative system. They knew the hard facts. They saw it coming.

But they managed to break the spell. They managed to put the all-pervasive megaphones on mute, and dance instead to the invisible music of an eternal tune. Enchanted, they overcame the odds. With the big pearl in their hands, they lacked nothing. Like King David or St. Francis, they ran around and danced naked, except for their boxer shorts.

And they laughed about it. Not with shallow, heartless laughter, though. Not with cynicism or self-indulgence. Not like tipsy ostriches digging their heads into the shaky sands of entertainment. They laughed with a joy that goes as deep as it gets. A strange wind had cleared the glittering fog, enabling them to imagine and breathe in alternative possibilities. The Floating Clowns were not bogged down by the heavy clouds of trends and stats, but rose above them like hot-air balloons instead.

And by living simply, having nothing and yet possessing all things, the Clowns became the world's laughing stock. Unlike the businessman who was so preoccupied accounting for all the stars, believing they were all his, the Clowns practiced an art forgotten by many. They smiled. They sang. They bent their heads backward taking in the wonder of the endless skies. They were filled with eternal bliss despite considering all the facts. Like Chesterton's angels, the Floating Clowns could fly feet-on-the-ground because they took themselves lightly.

Voilà. This outlandish mix of characters will guide us through the rapids of sex, Christianity, and climate change. But before moving ahead, allow me to lay down a few cards on deck as early and as honestly as I can.

Overcoming Our Roadblocks

Climate change and the ecological degradation our planet is going through are not glamorous issues. They go against the grain of our emotions; they call us to question our comfort; they require us to stretch our personal and collective dreams way ahead into a distant future; they make us realize that in the last two centuries, our societies have been feasting extravagantly, doped by the energetic steroids of cheap energy. German economist E. F. Schumacher's observation remains as true today as it did 45 years ago when he remarked that what our grandparents saw as a luxury, we now take for granted. That to say that leaving behind the lifestyle we've grown used to is no easy sell—especially realizing that around 80 percent of the world's energy consumption depends on fossil fuels. Speaking of an energy transition, which will require inevitable cutbacks, is like asking a teenager to take quick showers after 15 years of soaking in bubble baths.

In grappling with the ecological vital signs of planet Earth, it comes as no surprise that some of us will be tempted to dig our heads in the sand and ignore. However, if you are reading beyond what I just said, it's because you are somehow willing to grow nerves of steel and a heart of flesh that responds. Regardless, we won't be able to hide from our destiny for much longer, and denial, indifference, or inaction will not make things get better. Working for change does require determination, compassion, and lots of courage. It also requires the light-heartedness of the Floating Clowns—one that begins to grow in us when we're honest about important tensions within ourselves.

Getting Past Emotional Smokescreens

My own experience in dealing with the challenge of climate change leads me to realize we all have deeper emotional roadblocks that are often unspoken and that for the most part remain unresolved within us.[14] Consider at least three common reasons why acting on climate change has often been either dismissed or indefinitely postponed into the future.

Psychological. The first has to do with the mental stress that comes when our *view* of reality, of the divine, and of ourselves, fails to square with reality itself. Psychologists call this cognitive dissonance: the frustration (and denial) that takes place when we realize that the map we have been following is a faulty one, leading us to unwanted destinations. (A mouse will become a cat's dinner even if he believes there are no cats around. In such case his beliefs matter little; reality matters much.) This denial also comes to surface in what is known as confirmation bias, a tendency we all have to embrace information that supports our beliefs and reject information that contradicts them.[15] The apostle Paul, for example, once experienced confirmation bias when his outdated ethnic beliefs initially led him to be a proud persecutor of the early church. (Thankfully, he had an experience that made him change.)

One of these confirmation biases is now ingrained in our current economic system and the ideals it spreads, which, truth be told, many churches have sanctified and embraced. Ever since the Enlightenment, this system has made us believe that we humans are masters of our fate and owners of the planet. We have come to see the living world as an endless resource for our consumption, like a magic carpet under which we can hide our piles of waste. But Jeremiah, John the seer, and Boris Waterlove would call such a map into question, raising red flags to warn us about our ecological collision course, which we'll explore in chapter 2.

Is the map accurate?

Biological. Second, stands fear. Like every other animal, we have an instinct for self-preservation: we run away from danger, hunger, and death; we run toward safety, food, and life. Like every other animal, we spend most of our time thinking about our next bite, not about the meal we'll eat five or twenty years down the road. We shy away from difficulty; we praise convenience and comfort, now. And that's precisely what fossil fuels *have* brought us: the comfort and the ability to transport ourselves, grow our food, build more and more towers, and heat our spaces in a remarkably efficient way. Even though in industrialized nations we get around one unit of energy from our food for every seven units of energy we put into making it available, it's also the case that food has never been as convenient and accessible as it is now. "Why sacrifice taste, convenience, and low prices *today* at the vague altar of atmospheric stability *tomorrow*?" we may ask. It goes against our short-term instinct to question our comfort, even if that instinct is at odds with the comfort of those coming after us, including children, or even our own. And yet it seems that only a few are willing to float above normal by following the comical steps of the Floating Clowns.

Will we join them?

Relational. A final challenge emerges every time we face peer pressure to be like our buddies, workmates, or family members. Because it takes one person to carve a stone—but it takes a team to build a tower—it comes as no surprise that we have a deep-seated need for acceptance. Belonging and working in groups has always helped us thrive and survive. Collaboration makes us stronger. If we fit, if we're in, we're guaranteed a slice of the group's pie.

But what happens when our group—be it our family, our friends, our faith community, the organization that we work for, our nation—is following an outdated, faulty map that can't lead us into a destination we intend? We face the choice of either keeping a false sense of peace by comfortably going with the flow, or we can take courage and speak up about the need to change course. Silence and compliance can buy us belonging and acceptance, but being vocal can cost us our badge of membership. Most of us aim for the first because it feels good to be accepted by others in our group. However, what happens if our group is heading down a cliff?

Thankfully, not all people are rigid. Groups can be flexible and open to change. In fact, change *is* underway. Thousands of families have embraced a transit and biking culture; many "transition towns" are on their way to becoming food-sovereign and energy-independent; countries like Denmark, Germany, and Sweden are switching boldly to renewable energies; companies such as Interface are determined to become carbon negative.

Had the people of Jerusalem listened to Jeremiah and to all the line of prophets before him, perhaps they would have avoided paving their own road into exile. Today, time will tell if we'll be wise enough to learn from the past.

Will large groups pay close attention to the dissenting but enlightened voices crying out in the wilderness for ecological peace and renewal?

∼

Such are the two central convictions of this book: that ecological exile is knocking at our door (and now on a planetary level), and that personal and citywide repentance are necessary and possible. As will become clearer ahead, most of our modern cities and economies have been spiraling down, dancing their ways into endless cycles of adultery. We're still half-drunk on an intoxicating mix of fossil-fueled luxury and artificially-flavored entertainment. But we are still within a time-span of opportunity to avoid our fate becoming bound to that of Egypt, Babylon, and Rome, and all the great empires whose former glory eventually vanished into sand, dust, and ashes. There are increasing numbers of Boris Waterloves and Floating Clowns who have been paving the way of sustainability and the renewal of our cities, and it will be up to us to listen to them—and to change course not just cosmetically.

And to change course not just cosmetically, dare one underline, because our present-day ecological infidelities run deep. Just like Jeremiah spoke bluntly to his contemporaries back then, perhaps no one has put it more poignantly today than the above-mentioned American Christian writer Wendell Berry. "In denying the holiness of the body and of the so-called 'physical reality' of the world," Berry lamented, "modern Christianity has cut itself off from both nature and culture." His closing statement is worth quoting at length:

> Despite its protests to the contrary, modern Christianity has become willy-nilly the religion of the state and the economic status quo. Because it has been so exclusively dedicated to incanting anemic souls into heaven, it has, by a kind of ignorance, been made the tool of much earthly villainy. It has, for the most part, stood by silently, while a predatory economy has ravaged the world, destroyed its natural beauty and health, divided and plundered its human communities and households. It has flown the flag and chanted the slogans of empire. It has assumed with the economists that "economic forces" automatically work for good, and has assumed with the industrialists and militarists that technology determines history. It has assumed with almost everybody that "progress" is good, that it is good to be modern and up with the times. It has admired Caesar and comforted him in his depredations and defaults. But in its de facto alliance with Caesar, Christianity connives directly in the murder of Creation. For, in these days, Caesar is no longer a mere destroyer of armies, cities, and nations. He is a contradictor of the fundamental miracle of life. A part of the normal practice of his power is his willingness to destroy the world. He prays, he says, and churches everywhere compliantly pray with him. But he is praying to a God whose works he is prepared at any moment to destroy. What could be more wicked than that, or more mad?[16]

Like Jeremiah and John back then, Wendell Berry is a contemporary prophet who also has a big mouth—in this case addressing his words to the church in North America, sharing things perhaps too difficult for most of us to swallow.

Is he right? Are our cities complicit in delirious acts of ecological depravity, compromising the well-being of our earthly home?

2

———"Just Gimme the Facts"———
Ecological Vital Signs in a Snapshot

"The universe is made of stories, not of atoms."
—Muriel Rukeyser, American poet

"The goal is equality, as it is written:
'The one who gathered much did not have too much,
and the one who gathered little did not have too little.'"
—Paul of Tarsus (2 Cor 8:14)

CLIMATE CHANGE IS NOT a problem. Climate change is something else.
I realize this as I watch more than 2,500 miles of Canadian landscape run past my window. Departing from the West Coast, in Vancouver, the city where I have lived since 2008, the cross-country train in which I'm now sitting has been allowing me to witness two of the great wonders of North America—the Rocky Mountains and the vast prairies of Canada's interior. My amazement increases, realizing that we passengers are traveling comfortably inside a warm metallic bubble, enjoying oranges and bananas and two daily gourmet-style meals even as everything outside is dead cold well below the freezing point. (It's early February.)
This seems rather ironic for various reasons. Firstly, because the train ride would have been close to impossible without fossil fuels, and here I am

writing to express my concerns about the fact that climate change might actually be real and that it seems that it's we who are causing it. Yet it is also ironic because the train symbolizes one of the greatest achievements—and one of the greatest threats—of modern science and technology. It allows us to move from place to place in a matter of days or hours, but it drives an artificial wedge between us inside and everything else outside, even as it uses cheap fuels to release costly wastes in the form of greenhouse gases.

The irony escalates when one tries to figure out the Bible, arguably the most influential book of Western culture. Nothing close to an engineering or biology textbook, the Bible is an ancient, multilayered collection of sacred narratives, the first of which takes place in a garden. Far from being alien invaders, however, with poetic flare the narrator sketches an opening watercolor that depicts the representative of the hum*an* community as being one that is made from the hum*us*—a pre-scientific way of saying that we *soil*beings are somehow deeply dependent on the life that resides in the top*soil*. The narrator then shifts and focuses on the scene where humanity's representative is invited into a garden, to serve it and preserve it (Gen 2:15), not to abuse it or to sell it for profit. Eden is no enterprise to be managed, but a costly gift to be cared for.

Paradise Entirely Lost?

The story goes—and experience proves—that things went wrong in countless ways (Gen 3–11). But despite the ongoing acts of humanity's uncreative rebellion, the Most High remained committed to "all living creatures of every kind" and to "all life on the earth" (Gen 9:16–17). And so the treasured bundle of biblical narratives continues on, with the living source of the cosmos compassionately and yet fiercely determined to overcome human folly and selfishness with divine wisdom and life.

Contrary to popular belief, however, the Bible's vision for the cosmos does not end in heaven. Its final poetic brushstrokes portray a garden-city that is birthed out of heaven, so to speak, and that is coming down to earth. The drama resolves in a renewed and purified garden-city watered by a pristine river and fed abundantly by the tree of life planted in its middle—a 2,000-year-old vision of urban sustainability.

Tragically, over the course of the last several decades, some forms of teaching have made some Christians believe that the vision of the biblical authors sits rather easily with the earth being burned up and trashed to the side in exchange for a brand new planet (almost as if Yahweh was a twenty-first century shopping aficionado that shops and drops until a new earth corns and pops).

That belief is for the most part grounded in a misreading of a few sentences of the Second Letter of Peter, saying that "the present heavens and earth have been *reserved* for fire" (3:7a). As it turns out, such forms of teaching often fail to read the end of the sentence, where the author made clear that the fire was destined for the time of "the destruction of the *godless*." Right after, the letter speaks of "the elements being dissolved with fire" until everything done on the earth will be "laid bare" (3:10).

For the ancient Hebrew prophets, for Jesus, for Paul, and for the rest of the biblical writers, fire was always a *metaphor* for purification. By dissolving impurities, fire lays bare evil deeds but also discloses that which is pure and worthy and good. Fire was an image symbolizing the divine clean-up of everything and everyone that harms and pollutes the heavens and the earth—very much like a business owner filing a lawsuit and getting rid of those workers who steal and damage her property. The Second Letter of Peter says nothing about the extermination of the Earth for its own sake. As is evident from page 1, the biblical vision knows nothing about the Eternal One taking pleasure in destroying or burning up a cosmos that, from the get-go, the Eternal One declared "good" and "very good" multiple times. The issue lay elsewhere: John the seer and other writers of the Bible affirmed that the land and the living world are "groaning" hoping to be liberated from our human collaboration with the glittering forces of evil.[17]

Equally important, the biblical authors envisioned a paradise renewed. The final scene of the Apocalypse portrays a purified, Eden-like city; a sort of earthly icon of an eternal time when the streets will have no name and all the great achievements of cultures across time will be welcomed, propped up, and celebrated. With broad brushstrokes, John the seer portrayed the living source of the cosmos coming down to dwell among the entire community of life in an intoxicating feast of light and freedom (Rev 22; cf. Ps 148).

∼

It's easier for me to picture John's vision of cosmic ecstasy having just witnessed the radiant majesty of the Rockies, but the radiance makes me wonder: Should we bother to follow the steps of the Boris Waterloves who are looking after the wellbeing of planet Earth today—or is that something less-than-spiritual for tree-huggers and secular environmentalists to worry about? The chapters to follow sketch some additional reasons why the Hebraic-Christian tradition calls all of us to protect the integrity of the living world. But not before considering a few vital signs of the state of our earthly home—some of which are bound up to one of the biggest carbon bombs in North America: the Alberta Tar Sands.

Pipelines to Heaven

I write "carbon bomb" intentionally. The phrase is actually borrowed from an open letter issued by several thoughtful North Americans, including people such as former NASA scientist James Hansen and co-founder of the Land Institute, Wes Jackson, himself a Christian.[18] The document was written as a response to the Canadian government's consideration to develop the oil sands further, exporting voluminous quantities of bitumen through the proposed Keystone XL, Northern Gateway, Energy East, and Kinder Morgan expansion pipelines. (As I am writing this in my cross-Canada train ride, I am literally seeing hundreds of black, capsule-shaped train wagons go past me, shipping this heavy form of fossil fuel out West to China, very much as if they were mighty troops riding out of Mordor.)

I became aware of this letter in Fall 2014 following a summer-long research internship at A Rocha Canada, a faith-based conservation organization where I worked as a farm intern in the mornings and did research in the afternoons. Curious about the state of our planet, there I had the opportunity to understand better the reality of climate change and our highly-industrialized food systems. And perhaps nothing stirred my curiosity as much as interviewing Fred Bunnell, a professor that was visiting us from UBC.

"It's just unbelievable what we're doing to the planet," Bunnell said to me bewildered, struggling to find words. "It's just . . . it's just unbelievable . . ." Given he was chuckling as he played anxiously with his long white beard, at first I thought he was overstating it. But when I realized he's one of British Columbia's most prominent ecologists, his childish lack of words made me wonder if perhaps it's true that irony is a sadness that cannot cry but only smile. Whatever it was, Professor Bunnell's bittersweet laughter made me realize that climate change was not just a threat "out there," down sometime in the distant future.

"Just Gimme the Facts"

As I returned to Vancouver a few months later, some of the immediate implications of that conversation continued to become clear in an altogether different setting. A friend had convinced me to go visit a peaceful rally at Burnaby Mountain, the place where the Kinder Morgan oil pipeline expansion is intended to go through. On we went. Books to the side, feet on the ground, and long story short, upon our arrival the climate stats and figures crystalized in a rather tangible way that rainy November afternoon. With the memory of Bunnell's giggling face still haunting my mind, there we were now witnessing the Royal Canadian Mounted Police detaining dozens of seniors, academics, students, and well-meaning citizens who opposed the project. On behalf of a foreign oil corporation that managed to co-opt the Canadian state, the federal police were detaining Canadian people for protecting Canadian land and Canadian waters.

That alone should have been enough to make me consider the seriousness of Professor Bunnell's ineloquent statement. But despite the arrest of well-meaning, innocent people, and despite the fact that the voices of citizens and indigenous communities were ignored, the skeptic engineer in me still wanted to do some math—especially because I was taught that the cops were supposed to be the on the side of the good guys.

In brief, I realized that the heavy infrastructure of the Kinder Morgan pipeline expansion alone was set to increase daily crude exports from 300,000 daily barrels to 890,000 barrels per day, further unearthing Canada's proven fossil fuel reserves which amount to around 90 (and potentially 174) gigatons of embodied energy. But trippling the extraction speed seemed like pulling backwards, especially considering that four-fifths of those fossil fuels should stay in the ground if the nation is going to meet its goal of remaining below the politically-correct +2.0°C overall global warming limit—a limit that most scientists agree is nowhere close to being safe.[19]

It also didn't sound like an intelligent economic move. For one, a main-stream magazine such as *The Economist* estimated that total production of crude oil only represents less than 4 percent of Canada's GDP.[20] For another, even the Canadian Association of Petroleum Producers admitted that the oil sands supported the direct and indirect jobs of less than 0.65 percent of Canada's citizens.[21]

The oddity started to become clearer. Considering that dirty crude had decreasing economic returns and represented a meager slice of the nation's GDP; seeing how native communities and their lands in Alberta and British Columbia were bullied by corporate giants (and continue to be to this day); knowing that shipping bitumen oil was and is a threat to Vancouver's beaches and its watershed; aware of how the energy trapped in the oil sands is a "carbon bomb" to the global climate; witnessing how the pipeline led to the arrest of seniors, scholars, scientists, and innocent people . . . the numbers didn't square all that well.

While more progressive nations like Germany and Norway have been taking great steps toward decarbonizing their economy, some amongst Canada's elected politicians have continued to this day to do their best to move their nation in the opposite direction.

As outlined in the Annex, this mismatch became the spark that ignited Earthkeepers into existence. Jeremiah, John the seer, and several Boris Waterloves inspired a few friends and I to kick-start a faith-based initiative aiming to partner with other organizations addressing some of these issues. We felt something was missing and that churches remained largely unaware (or dead silent) about the deeper dynamics of the challenge. So having witnessed some of the power imbalances around the pipeline decision, a few of us decided to open a conversation: Were scientists right in affirming that climate change was as bad as they claim it is? And should people of faith respond to what scientists are saying?

Science or Religion, or Science and Religion?

Copernicus and Kepler. Newton and Pasteur. Mendel and Plank. These were all accomplished scientists with a faith in God. Even Einstein, who himself was likely not a theist, remarked that he "wanted to know how God created the world" and that he wanted "to know His thoughts" because "the rest are details."[22]

For over 150 years, however, science and Christianity have been at great odds. Many scientists in the West (and engineers, like myself) have been taught that all that happens can be entirely attributed to natural causes, that so-called miracles are fantasy stories, that faith is blind, subjective, or sentimental. We have been taught to assume as fact that God is an illusion and to have faith, instead, in the belief that matter is all there is.

Threatened by these claims, some people of faith have gone to an opposite end: despite the good intentions, some devout Christians today read the Bible as if it was a book of modern science and naked facts. Many believers are convinced from the get-go that the good old book *must* answer the kinds of scientific and historical questions we modern people are so eager to explore (in how many exact days did the universe come into existence, what's the chemical "stuff" that we're made of, etc.). Influenced by the materialism of modern science, this perspective tries to read as fact what is often meant to be read as metaphor and poetry—which would mean that Jesus would have had actual branches growing off his ears when he affirmed he was "the true vine" (John 15:1), or that the heavens and the earth were created in exactly six twenty-four-hour days.[23]

But are dry facts and cold stats what the Bible is actually about? Or is it something far more interesting? Conversely, can science easily dismiss the Bible as being childish and unscientific?

This split between science and faith has left us with two camps, often misunderstanding and fighting each other. Many naturalists reject Christianity as ancient superstition; many Christians either ignore secular science or distort the Scriptures by wanting to turn them into a biology or astronomy textbook. In doing so, both sides argue infinitely, very much like people using two very different sorts of maps to navigate the same landscape. Scientists feel frustrated because some readers of the Bible read the book as a political map to find answers, say, to geographical questions. Upset, some scientists fight back, convinced that geographical maps are used to describe political boundaries.[24] Given the lack of agreement on basic terms, the battle goes on without end.

But who's right?

Both are. But both are wrong too. The two sides often fail to realize that, on the one hand, modern science attempts to answer questions about *how* we got here and *how* things work. On the other hand, the Bible's first eleven chapters are an ancient cosmological account of *why* we exist, in *whose world* we live in, and *what our role is* in the entire drama. Science seeks to describe what *is*, religion illuminates what *should* be and for what *purpose*. Science and religion are like two different kinds of maps that are meant to describe different dimensions of the same reality.

This match between some scientists and some believers is puzzling as well because the origins of modern science were in fact considerably rooted in Christianity. As shown by Oxford historian Alister McGrath, the motivation and many of the intellectual foundations of modern science were laid down once European scholars and scientists began to take for granted that the divine wisdom of a good Creator was manifested in the order of the created world. As the physical world came to be understood as a good place brought into existence by an uncapricious, good God, folks gradually realized it deserved to be seen, touched, and studied. The study of stuff would reveal something about the mind of God. Western science in the seventeenth century came to recognize that studying the material world was important because matter *mattered* to God, and because it reflected the beauty and wisdom of the living source who authored it and sustained it.[25] (In that sense, one could even affirm that God is the first and greatest materialist.)

All this to say that scientific study should matter deeply to those who confess God as the origin and living source of all that is. The main interest here, however, is not convincing either scientists or theologians to shake hands and be in dialogue—as they should. Rather, the rest of this chapter sketches a few of the ecological vital signs that science has been disclosing about the state of the living world we all share. (Less analytically inclined readers might want to jump straight to chapter 3 and following.)[26]

Is Climate Change Something New?

When it comes to issues like climate change, some recent claims want to make us believe that we live in a post-fact world. Other politicians argue for alternative facts. Still others call us to remain hopeful after we have considered all the facts. And even others say that facts by themselves don't spur us into action, but more likely into despair or denial. I agree with the latter, and so do the Counting Crows and Joni Mitchell: Simply knowing the facts has not kept us from paving paradise to set up our parking lots.

Still, facts and estimations have a place, and it takes courage and sobriety to face them. Despite their ongoing disagreements after several decades, today the world's leading scientists have been arriving at a fairly unified common voice—a scientific consensus. Drawing on their findings, we can have enough certainty of what has happened, and what is likely to happen, in the coming decades because of climate change.

Whether primarily induced by humans or not, the stability of the atmosphere is already at risk, and we'll only ignore it at our peril because the relative balance of the Earth's climate and its regular seasons have been critical for the recent success of our human species. A relatively stable climate is what has allowed for the relatively steady food supply that has come through agriculture. And that, in turn, has been the basis for the rise of the Mesopotamian, European, Inca, Chinese, and the twenty-odd different civilizations that have flourished around the globe since the end of the last Ice Age between 11,000 and 13,000 years ago.[27]

The math is surprisingly simple: Stable climate, food, civilization. Strikingly, so is a question: Will the millenary stability of the climate endure?

Atmospheric Powers, Unleashed

To be sure, the Earth has gone through at least five periods of mass species extinction before we *Homo sapiens* claimed center stage. As it turns out, some of these extinctions took place at least partially because the climate *has* changed many times in the Earth's past without any human intervention. Shifts in solar irradiance and volcanic dust have always contributed to temperature fluctuations in the atmosphere and the oceans, for example. In fact, more than 2,500 scientists from the Intergovernmental Panel for Climate Change (IPCC) have very high confidence that between 129,000 to 116,000 years ago, the world's mean sea level was at least 5 meters (16 feet) higher than it is today. Because of warmer temperatures during the last interglacial period, sections of the Greenland ice sheet melted, very likely contributing between 1.4 and 4.3 meters to the higher global average sea level.[28]

In our day and age, however, changes in total solar radiation are estimated to have accounted for only around 2 percent of the total warming sources in 2011, relative to 1750 (the decade when the coal-powered Industrial Revolution began to take hold). In what is now its fifth official report, the IPCC scientists have shown how the years from 1983 to 2012 have been the warmest thirty-year period in the past 800 years. Their studies continue to show a linear trend of average temperature rise of 1.53°F/0.85°C starting in 1880. In turn, from 2000 to 2010 greenhouse gas emissions were the highest in human history, reaching concentrations of carbon dioxide, methane, and nitrous oxide to unprecedented levels in (at least) the last 800,000 years. In turn, these findings reveal that CO_2 emissions from fossil fuel combustion and human industrial processes have contributed about 78 percent to the total GHG emissions increase between 1970 and 2010. According to the IPCC, this suggests that it is "extremely likely" that we humans are the dominant cause behind this particular warming period.[29]

Not all this is doom and gloom. Fossil fuels *have* made life easier and they have helped to lift millions out of extreme poverty. In many cases, higher levels of CO2 have also led to higher crop yields and higher levels of plant growth. There is no doubt that burning these natural fuels has served us well in more ways than one. What is striking, however, is the pace and the intensity of these changes of temperature in the atmosphere and the oceans, which have little precedent now that we have entered the Anthropocene—an era in which human cultures and technologies have become the predominant force in determining the fate of the planet's ecosystems. In a way, it could be said that it has taken us 250 years at large, and around 50 years in particular, to undo the longstanding work of natural cycles that have very gradually captured and stored energy from the sun to cool down our earthly home to a humanly-hospitable level. This can be pictured differently: If these 800,000 years were squished into a twenty-four-hour time scale, we've burnt most of our fireworks and drunk our carbon cocktails in little less than 6 seconds.[30]

Put simply, the heat is on, and it's now leading us to face the consequences of going past our curfew. Truth be told, Yale's 2018 Environmental Performance Index report indicates that three-fifths of countries have slightly declining intensities of CO2 emissions—a trend which the report identifies as promising but that must nevertheless be accelerated.[31] For one, in 2016 the Institute for Health Metrics and Evaluation estimated that diseases related to airborne pollutants contributed to two-thirds of all life-years lost to environmentally related deaths and disabilities—problems especially acute in swiftly industrializing and urbanizing nations such as India and China. For another, rising levels of CO2 means a greater likelihood of more forest fires like the one in 2016 in Fort McMurray, Alberta, or 2017's in British Columbia or California. More CO2 intensifies the aggravating affect El Niño has on places like Portugal and Indonesia (whose forest fires alone, for instance, were estimated to have released between 13 to 40 percent of the entire world's annual fossil fuel emissions in 1997).[32] In times past, there used to be extended periods between hurricanes the size of Harvey, Irma, and Maria. But our warmer skies, which happen to hold more water, mean these hurricanes are now striking us more often. More CO2 means a compromised water supply and more failed crops in Mexico and California (elevating food prices in the US, Canada, and elsewhere); it also means more propensity for urban floods like that of 2013 which displaced over 250,000 people in Calgary; more Sandys, Ottos, and Othelias whose tragic outcomes are too well known.

Lest we're led to think otherwise, significant changes to the climate are already underway, and our adaptation must occur. But unless highly industrialized nations change course quickly, the future will be no brighter. In the four possible future scenarios identified by the IPCC, the average surface temperature increases will range from 1.8°F to 9°F (1°C to 5°C) by the end of the twenty-first century. This makes it is very likely that heat waves will occur more often and last longer; that wheat and corn crops will decrease their yields; that extreme precipitation events will become more intense and frequent. In short, moodier Katrinas, stronger Friedas, and drier Californias.

Troubled Waters Ahead

Interestingly, only around 1 percent of this additional heat is stored in the atmosphere itself. Most of it is in the oceans, which currently hold an estimated 90 percent of the energy increase accumulated between 1971 and 2010. This is no small thing, because temperatures changes in the seas means disturbances in the normal cycles of ocean and air currents. The excess energy creates a greater propensity toward abrupt movement, randomly affecting precipitation patterns that have been more or less stable over the centuries.

For one, the added heat has also brought an enlargement of the oceans, given that water expands when it's heated. But besides the expansion, the IPCC shows how the melting of glaciers from the Greenland and Antarctic ice sheets, and of land water stored in the form of snow and ice, is connected to 75 percent of the observed global sea level rise since the early 1970s. Starting in the 1990s, for instance, the decrease rate of the Greenland and Antarctic ice sheets has been in the range of 3.5 to 4.1 percent per decade.

Both the ocean's enlargement and the water from melted glaciers has meant that, during the period from 1901 to 2010, the global annual sea level rose, on average, 20 cm. That may sound trivial, but the pace is faster than the mean rate during the previous 2,000 years. Glaciers have continued to shrink worldwide, even as the Northern Hemisphere's spring snow cover remains on the decline. The IPCC's range of estimations points out that by 2100, sea levels may increase by 0.4 to 1.0 meters, while glacier volume is likely to decrease anywhere between 15 percent to 85 percent, affecting what has otherwise been a fairly steady water supply to hundreds of regions around the world. (Glaciers and snow caps function as massive coolers that allow the snow to melt slowly, permitting it to flow gradually downhill, instead of running off all at once.)

Changes like these could likely affect over 200 million people worldwide in coastal regions, increasing the likelihood of floods, as well as stressing the summer water supply of areas that depend on streams and rivers for irrigation.[33] More broadly, the International Organization for Migration affirms that the sum of climate-led effects could likely translate into an estimated one billion ecological refugees by 2050.[34]

Likewise, since the beginning of the industrial era, this uptake of CO_2 by the oceans has resulted in their waters shifting toward becoming acidic. With more hydrogen ion concentration, the pH levels of ocean surface water have decreased by 0.1, corresponding to a 26 percent increase in acidity. Regarding its magnitude and rate, this observed trend of CO_2 concentrations in the oceans is unparalleled in the oceans' history of the past 20 million years. Such low pH values in seawater hurt the formation of carbonate minerals, which is critical for many invertebrate marine animals with carbonate skeletons, such as mussels, corals, and sea urchins. And while mobile and active beings such as fish and certain crustaceans are much more CO_2-tolerant, it's also the case that warmer waters have less capacity to hold oxygen, thus constraining fish habitat. Some marine species will undoubtedly adapt to this change better than others will, but what is disconcerting are the yet-unknown ways in which these hyper-fast increments toward acidity may (or may not) disrupt the ocean food chain, founded as it is on microscopic marine phytoplankton.[35]

Next to these changes, one must also mention the well-known flooding of the seas with an epidemic of petroleum-based plastics; the bodies of water poisoned with mercury, reaching into the lifeblood of countless creatures; the piles of toxic electronic trash exported to the far East; the genetic blending of the DNA of bacteria and plant species; the increasing decimation of pollinator bees; the monopolization of modified seeds in the hands of corporations claiming to be as wise as gods. And so on. As our technological society advances under the highly-applauded dictatorship of growth, the earthly paradise and its hospitable climate continue to become anything but desirable.

"Just Gimme the Facts"
Related Signs & Unfading Facts

It's easy to forget that every ecosystem is a food system where life and energy flow from one creature to another: plants transform the sun's energy, animals eat plants, animals die, they all decompose and feed the soil, and so on. That's basically why an ecosystem collapse means the collapse of the food web that depends on it.

Many giant agribusinesses seem to ignore this. In the 1970s, the so-called Green Revolution brought heavy machinery into the fields to double outputs of corn, wheat, and rice between the 1950s and 1990s. To do so, however, it has since relied on the intensive use of synthetic fertilizers and pesticides, which themselves require fossil fuels to be produced, plus the heavy equipment to grow, process, transport, and store food. All in all, industrialized food systems are using up around seven units of fossil-fuel energy to grow one unit of food energy (leading some to lament the fact that we can't eat crude oil).

Great risks have come with such fossil-fueled transmutation of small, diversified farms into industrialized monocrops controlled by Star-Wars-like machinery. Our seemingly endless urban sprawl and our unsustainable tilling and irrigation practices have been eroding the soils and ecosystems that sustain us. As of today, we have cut down around one-third of the world's forests. The UN estimates that already an area roughly the size of China suffers from recent desertification (an area that could otherwise supply 20 percent of global food production). Of the three percent of the water that is drinkable, two percent remains locked up in glaciers (which are shrinking because of warmer global temperatures), and only one percent is fit for human consumption. Large-scale agriculture requires 70 percent of such water; but by using it willy-nilly, the food industry continues to deplete aquifers, even as its heavy irrigation techniques increase the salinity (and infertility) of soils. In turn, the human demand for fresh water is estimated to double by 2050, but already by 2025, 65 percent of the world's people will likely suffer water stress.[36]

Like Boris Waterlove, many of us find it disturbing to consider these predictions—ones which are closely intertwined with a series of other consequences. Along with the dire fate of an increasing number of vulnerable poor human populations, our fellow non-human creatures are also at the forefront of the struggle. As outlined in the WWF's 2016 Living Planet Report, species populations of vertebrate animals around the globe have decreased by 58 percent between 1970 and 2012, at a pace 1,000 times faster than normal, mostly because of overexploitation and the loss and degradation we have caused to their habitats. (One could picture this by imagining an extra-terrestrial species invading our planet with destructive spaceships, suddenly forcing two-thirds of us humans to disappear, even as the aliens claim to have the right to eternal life, freedom, and the pursuit of happiness.) Still, even if climate change is not the sole or main culprit behind this mass extinction, it has and will certainly exacerbate the harm. Abnormal and shifting temperatures will affect reproduction and migration patterns, as well as causing uncontrollable outbursts of mosquitoes and new tropical pests in previously unaffected areas, such as the south of the United States.[37]

Our civilization is a remarkable achievement in countless ways that don't need to be championed here. Still, our effects on the living ecosystems have made climate change a matter of our survival and that of all other living beings. In fact, the extent of extermination approximates that of the last major extinction which eliminated the dinosaurs, with one exception—it's occurring way faster and we are not exempt from this threat. A recent study from the National Academy of Sciences of the USA concludes:

> The resulting biological annihilation obviously will have serious ecological, economic, and social consequences. Humanity will eventually pay a very high price for the decimation of the only assemblage of life that we know of in the universe.[38]

∼

So say the scientists. And this brief sketch makes one wonder if the big mouth of Jeremiah would have labeled this as a global emergency set wild by lascivious fossil-fueled intercourse—one that shows little signs of being truly sacred.

A Parody of the Garden-City

Our glitzy and ready-made supermarket culture is an artificial buffer that makes us rather unaware to our ecological emergency. It also makes us forget that civilizations are little less than big cities with hungry bellies. Cities require a constant food supply, and food supply requires stable irrigation, healthy soils, and predictable weather patterns. Stable civilizations require no extreme floods or droughts or frozen crops. On the flip side, unstable climate means less food, and less food is a red flag for civilization. As dumb as it sounds, it all comes back to basics: climate, food, civilization.

To be sure, our 10,000-year-old experiment of building human settlements and cities is no small feat. Building on achievement after achievement, we have created outstanding medical devices that extend the span of our lives and countless technological gadgets to make them easier. We compose love songs and sublime symphonies; we celebrate with wine and dance and fireworks. We collaborate to build hospitals, and schools, and outer-space telescopes, and a global artificial brain we call the internet. We humans have colonized and mastered areas of the planet that for countless centuries remained significantly hostile to human habitation. The collective achievements of modern civilizations are nothing short of remarkable, in countless ways. Wealth and material progress have certainly spared billions of us from many of the harsh, unpredictable perils of the bush. It thus comes as no surprise that the 2015 World Happiness Report identified income levels, relational support networks, and healthy life expectancy as the three most important factors linked to human well-being. Having enough money can buy us a good measure of well-being.[39]

But will our heads shine only for the time being, at the expense of the future of those coming after us, or even our own? Our ecological track record proves that in signing off the contract we have come to think too highly of our ourselves, often to the point of bypassing the small print—a small print perhaps too unpopular for most to consider, let alone to bear or confront its implications.

The Burden of Proof Falls on Hungry Cities

What are some of the hidden costs (or "externalities," as economists call them) that allowed my fellow passengers and myself to enjoy a banana from Ecuador as we blazed across the world's second-largest country in a diesel-powered train? What's the cost of paving paradise to set up our parking lots?

Far from informing mainstream business practices, a motley crew of eclectic ecological economists has been giving us insights into these questions. "We're living in planetary 'overshoot,'" they tell us. The more poetic among them say we're sawing off the branch on which we're sitting; that we're tearing down the wooden walls of a handcrafted house to feed a fireplace to warm us up. The more factual among them pin this down to 1.6

planets: that is, we currently need 1 2/3 Earths to sustain current levels of human consumption and to absorb current levels of human waste (mostly in the form of greenhouse gas emissions). Hence the "overshoot."

Consider the main city in the unceded territories of the Salish Sea in British Columbia, where I had the opportunity to live from 2008 to 2016. For one, Vancouver's ecological footprint is between 4 to 7 global hectares (gha)—a measure between two and four times the size of the global ecological fair share, which is around 1.7 gha. (Global hectares measures the amount of land needed to plant trees that absorb carbon emissions, as well as the amount of space used for fisheries, agriculture, construction, water use, etc.) Interestingly, around 52 percent of those global hectares are what ecologists call "forest land," the space required to plant trees to absorb Vancouver's greenhouse gases. In other words, half of the total footprint used up by Vancouver is needed to compensate for its pollution of the atmosphere.[40]

This is to illustrate that if humanity were to stand before a planetary judge, the burden of responsibility would fall on the modern-day cities—the emitters of up to 80 percent of the world's greenhouse gases (aka GHG's).[41] Glamorous as they are, our cities are an achievement nowhere close to being possible without the invaluable assistance of over 250 years of cheap fossil fuels. Building skyscrapers, growing and moving food around, heating our living spaces, traveling to and from work—all this and more has been possible because fossil fuels have been readily available. But what if we stopped taking them for granted, as we should?

A quick review of basic thermodynamics might be of the essence. For one, fossil fuels are nothing but compacted solar energy transformed and tightly packed up in the form of decomposed organic matter. For another, scientists remind us that everything in nature obeys the law of entropy. This law dictates that without maintenance and a continual flow of energy, everything tends toward chaos. Everything. Rooms get dusty, clothes wear out, roads crack up, atmospheres get wild. We find order in our world only because the earth's living sphere is empowered by the sun, whose energy transforms elements into ordered biological matter (plants, trees, wood, animals, humans... and eventually fossil fuels). If it weren't for the sun's vitality, things on earth would follow their natural tendency toward randomness and disorder. No energy, no order; no sun, no ecosphere.

And here's the rub: Cities are hubs that disperse that finite energy, which is so tightly packed up in fossil fuels. Cities (like organisms) can only maintain order at the expense of creating *dis*order somewhere else. To put it candidly: ordered, well-fed cities also need to release their farts and dump their poop somewhere else—literally and figuratively.

Today, our artificial separation from living ecosystems that sustain us makes many of us blind to this reality. Harsh as it sounds, however, cities are essentially parasites sucking up land and water ecosystems to sustain our global population, now of billions, and then export our waste. Ecological economists don't exaggerate in saying that cities are true "engines of ecological decay"—they flourish at their center but use up and degrade whatever is out of sight. While they occupy only 2 to 3 percent of our planet's land area, the ecological footprint of some cities requires an area up to 250 or 300 times larger to sustain them (land to grow food, feed animals, plant trees to absorb CO_2, etc.).[42]

Take the case of greenhouse gas emissions. For the earth to assimilate our burning of fossil fuels (by absorbing and locking up the CO_2 into soils, plants, and trees), each person's individual lifestyle would need to release no more than 1.5 tons of GHGs per year. However, high-income countries like the United States, Australia, and Canada have emissions of over 19 tons per person per year, equal to each one of us burning over 1,900 gallons of gasoline every year by driving over 42,000 miles.[43] As it turns out, the wealthiest 20 percent of the global population stands behind over 75 percent of private consumption and its associated pollution. A small privileged section of humankind is using up the earth's entire bio-capacity. (Unfair) overshoot indeed.[44]

Cultural historian Ronald Wright nailed it in his now-famous lectures on *A Short History of Progress*, highlighting how our fossil-fueled, swift technological wizardry has led us to become, as he puts it, "too successful." (Or, to be precise, a small *fraction* of the population is enjoying the party at the expense of the vast majority of the human and non-human world.) By creating cities and economies that allow us to deplete soils, forests, and all the renewable gifts of creation at a faster rate than they can be replenished, we are locking ourselves up in what Wright calls "progress traps." The very means that have allowed industrial civilizations to flourish (fossil fuels) are now putting our own existence at risk.[45]

Wright is not alone. Prominent urban ecologists Jenny Moore and Bill Rees are bolder in their diagnosis: "Humanity has become dangerously parasitic on its planetary host."[46] And Patagonia's CEO Yvon Chouinard is perhaps more vocal when it comes to describing what he calls "the old economy" when he says, "It's as though we'd handed Satan a hard hat and asked him to refashion our earth according to his plan."⁴⁷ In turn, realities like these have given Pulitzer-price journalist Elizabeth Kokbert enough reasons to write *The Sixth Extinction: An Unnatural History*—a book whose title speaks by itself when it comes to pinpointing the essence of our shared challenge.

Ask whom we may, it's clear that the consequences of inflicting such harm on the earth's ecosystems are expected to grow over time, increasing food and water insecurity, cracker-crumb housing, obesity and heart disease, raising prices for many goods and services, as well as conflict and migration. The harm is also turning a beautiful paradise into a stressful dwelling place covered by an ugly sadness.

Still, we are tempted to doubt or dismiss the possibility that our global civilization is in fact in jeopardy. "Are we humans responsible?" "Are we being 'so arrogant,'" as some proponents of alternative facts are claiming, "as to think that we can actually *change* the entire planet's destiny?"

In truth, we already have. Our misguided and unrestrained desires have created an entirely new climate.

∼

Climate change is surely not a problem. Climate change is a *symptom* of many problems—one of which is our one-way ecological love affair propelled by the endless extraction and the mindless burning of fossil fuels.

To those of us who are yet urged to safeguard the future of others as if it was ours, this challenge likely sounds trivial and distant. It surely is tempting to join the ancient Corinthians in saying "let's eat and drink for tomorrow we die" (1 Cor 15:32). One could also cheer for a former colleague of mine who once admitted that "whatever happens to two or three generations down the road . . . doesn't really matter." Living for the belly, like the Corinthians, and remaining indifferent, like my colleague, are two ways of coping with the challenge. We can surely dig our heads into the sand and live without a sense of responsibility to others, nor to the One whom all three Abrahamic faiths acknowledge as the living source and owner of everything that is.

Another way to engage with our ecological challenges is by learning from the past—both to avoid the same mistakes but also to take inspiration from others who have gone before us. In *Collapse: How Societies Choose to Fail or Succeed*, UCLA biogeographer Jared Diamond contrasted short-sighted societies with wiser ones. Some, like the Easter Islanders or the Greenland Norse, were enchanted, cutting down trees to build statue after statue but then failing to reforest their lands, eventually eroding their way into exile (or into cannibalism). Others, like the Tokugawa in Japan, saw it coming and, among other shifts, managed instead to set up timber quotas to limit consumption and protect their island against landslides and timber depletion. Still others in more modern times have established strong regulation to reduce ill health effects—as has been the case of sustained efforts by the US government since the 1970s, which eventually cut down 6 major pollutants by 25 percent (even as the US population increased by over 40 percent in that same period of time).[48]

It's often said that when history repeats itself, the price goes up. Ignoring what went on before us makes us prone to falling into the same foxholes; remembering it can be a source of wisdom, and, sometimes, of inspiration. Having sketched a few of today's ecological vital signs that scientists have been exposing us to, we'll now revisit our history through the eyes of faith by revisiting the ancient story of Babel and Babylonia—one which, as we'll see, has been recently remixed in California.

3

Babel, Babylonia, & California
An Ancient Story, Remixed

> "We're more popular than Jesus now;
> I don't know which will go first—rock 'n' roll or Christianity."
> —John Lennon

> "When I give food to the poor, they call me a saint.
> When I ask why the poor have no food, they call me a communist."
> —Hélder Câmara, Brazilian Clergyman

CONSIDER AN ADVERTISEMENT OF a blue jeans brand. With a picture of a semi-naked body in the background, the billboard read: "Smart listens to the head, stupid listens to the heart." It was finished off with a "Be Stupid" slogan in the bottom-right corner. Or take the case of Apple. "What iPod are you?" we're asked, likely suggesting a new spin-off on Descartes' famous statement, now implicitly reframed as "iPod therefore I am." To this, Visa's clever ads constantly urge us to "Go," and McDonald's follows suit with its "I'm lovin' it" jingle, laid over an image of a (very) foggy car with the haze in one of its windows wiped by a passionate five-fingered tear-off: Fast sex for a fast food world.

These are a few of the many megaphones clamoring for our attention today. For one, they push us into a me-centered treadmill, an unending "i"-world of Go-Push-Go-Enjoy-Here-Now. For another, they try to empty our pockets by luring us into being (consciously) idiotic. (*Rather smart, isn't it?*) And while a few progressive enterprises have taken a stance by drastically cutting down their advertising expenses, most of today's organizations devote a good slice of their budget to all things marketing. Standard business theory knows all too well that advertising is the art of endless repetition, of creating dissatisfaction, of tweaking our desires to keep us coming back for more.

To all this, however, most of us remain like fish in a stagnant pond, not knowing what it feels like to be wet: swimming in the murky waters of ads and promos is all we know. Even if the living world continues to crumble at our feet, most of us are tempted to grow blind to these voices. We don't know any differently. More disturbingly, some have taken the megaphones for granted and seem to be rather okay about consenting to their call. "Be Stupid." "I'm loving it."

Is this normal? Is this i-want, fast-food world simply the way things are and have always been?

It's certainly ambitious to summarize in a few pages a story that takes bookshelves to narrate. Because space is of the essence, a brief draft will have to do in unveiling some of the main forces that have led us into such a crazy little thing called shopping love. The sketch below will at least begin to give us a hint as to how this colorful consumerist cocktail relates to sex and Christianity and climate change—and what that has to do with Babel, Babylonia, and California.

Sketching the (Civilized) Story Behind the Facts

Human extermination of other species and human harm of the living environment that sustains us is nothing new. In his now famous *Guns, Germs, and Steel: The Fates of Human Societies*, the above-mentioned UCLA biogeographer Jared Diamond recalled how the earliest human populations have always altered their ecosystems, sometimes to the point of their own downfall.

Consider again the Polynesians of Easter Island who ended up cutting all their trees to roll their big stone statues. Doing so led to exposing their agricultural soils to erosion. It also meant no more timber to build canoes, which then resulted in less protein from fishing. The tools they used, the myths they believed, and collective habits they practiced enabled their population to become greater than the island's living capacity could support—even if that eventually led their society into a holocaust of warfare and cannibalism. By failing to care after their earthly nest, the Easter Islanders paved their way into their own ecological exile. Likewise, the Chaco *pueblos* in what is now New Mexico once lived surrounded by woodland in what they eventually turned into a desert. They grew into a relatively advanced civilization, but deforestation meant water shortages, and eventually the abandonment of what today are the ruins of former five-story-high dwellings. A similar fate awaited the Maya in Central America, the Harappan in India, the Greeks, Romans, and Persians in the Mediterranean: each flourished for a time but only at the expense of eventually depleting their resource base through overgrazing, erosion, or soil salination. The fruit gardens of old were not always the empty corner stores they are today. "The ruins of those cities," as Diamond puts it, "are monuments to states that destroyed their means of survival."[49]

Early humans also exterminated large animal species. Thousands of years before the European colonization, the first settlers in what is now Australia/New Guinea cleared out 400-pound ostrich-like birds, one-ton lizards, giant pythons, and land-dwelling crocodiles. In the Grand Canyon area, too, aboriginal humans wiped out ground sloths, mountain goats, mammoths. And likewise around the globe, from the strait of Gibraltar to the banks of the Huangpu River in Shanghai, from the westernmost tip of Alaska to Tierra del Fuego and La Patagonia. The worldwide occupation of *Homo sapiens* and the eventual rise of relatively advanced human settlements has meant the decline of much undomesticated mega-fauna. The Industrial Revolution was clearly not the first ecological parasite sucking the life out of what we often suppose to have been an otherwise pristine natural paradise.[50]

What *is* new to our day, however, is the magnitude and the pace with which Westernized cultures and our economic systems have been causing the damage. Because of their loss of habitat due to our unsettling human impact, today non-human creatures are currently disappearing at a rate 1,000 times faster than normal. Or seen differently, it took our hunting and gathering ancestors around 900,000 years to move away from crude stone axes to less rudimentary rope nets, spear points, and fishhooks; it has taken us roughly twenty to flood the world with smartphones, threatening to send landlines into the creatively destructive cemetery of obsolescence.

Our "progress" has sped up to the point of inflicting great wounds on all life around us.[51]

How did we get to this point? What's the story behind the rise of cities, consumerism, and climate change?

Away from the Land, Into the Cities

The earliest settlements and cities began to emerge as far as 11,000 or 12,000 years ago (*hello cities!*). This happened partially because our ancestors were able to tap into previous innovations, such as the use and development of more complex languages, the controlled use of fire, and the passing on of technological knowledge around the use of rudimentary hand tools. Still, the leap that led to the rise of civilization took place mainly because our species learned how to get a grip on a more steady and relatively constant source of energy. Instead of wandering around hunting and gathering, folk in times past began to sow seeds and domesticate a handful of creatures. When the Agricultural Revolution was born, it gave our ancestors an unprecedented ability to settle in one place—and thus begin to deplete virgin, non-renewable materials at a growing rate. Likely too gradually to even be noticed, ever since then, we humans have gone from living a "passive solar existence" of hunting and gathering, to a life based on constantly extracting energy from the earth's "carbon pools"—namely soil and wood, and eventually coal, oil, gas, and more recently, bitumen.[52]

Before agriculture, early humans could travel everywhere, because they traveled light. They pulled out starchy roots for nourishment, picked berries for delight, killed animals for fur, and burned fallen branches for heat. But when settlements, and then cities, arrived on the scene in the Ancient Near East, and eventually elsewhere, the traveling load became rather heavy. Our ancestors didn't find it very convenient to transport temples or palaces nor their unprecedented surplus of food, which itself required building places for storage. As opposed to moving around chasing after her renewable gifts, settling in nature seemed rather convenient—even if that would gradually lead to grabbing her by the throat.

The increasing flux of gadgets and cultural achievements overshadowed the seizure, so the downsides of civilization surfaced relatively slowly to be immediately perceived. In fact, despite being prone to deforestation, as in Easter Island, or to the salinization of fertile soils, as in Sumer, settlements and cities had a lure of their own. Depleting an out-of-sight soil patch from its nutrients went unnoticed (at least at first); sculpting the Great Sphynx in Egypt's old kingdom or lighting fireworks in medieval China, did not. The glamour of power and spectacle has always pushed farmers and slaves toward the bottom of the pyramids. Surely buildings are much more attractive than soils; even if the first cannot exist without the constant nourishment and food supply that come from the second.

Still, from their early emergence until around the 1500s, cities were home to only 5 to 10 percent of the entire human population (whereas today they host approximately 50 percent). Early villages and settlements had minor ecological ripple effects that only sent wide, dissonant soundwaves whose noise traveled relatively unheard. But when cities reached their puberty, the dissonance began to be heard afar. And nowhere else did the noise grow as loud as when folk in England started to use coal to warm their houses—and eventually to power their factories a few centuries later. Home to perhaps 50,000 inhabitants, by the 1500s, London was smaller than Beijing. But come the 1900s, the capital of the British Empire became a global megalopolis bursting with 6.5 million people. What was the hormone that unleashed the boost?

To be sure, Western Europe's plundering of the Americas led to a flow of gold and silver that set the base for the accumulation of capital necessary for the continent's upcoming industrialization. Shipbuilding, gun foundries, and other imperial endeavors were to an important extent financed with booty and foreign treasures. "The Industrial Revolution," highlights Ronald Wright, "begins with Atahuallpa's gold." The Revolution was then aided by shipping African slaves to European colonies, forcing them to grow cheap cotton, sugar, and coffee for European cities.[53]

Despite its central role, imperialism accounts for only half of the story. Accounting for the eventual rise of the West above other cultures, Harvard historian Niall Ferguson attributed Western domination to science, private property rights, literacy, a disciplined and thrifty work ethic, competition, and (*tada!*) consumerism. And that's all true, for these were all key ingredients that gave the West an edge over the rest. Come the Renaissance, Western Europe was, in Ferguson's words, a "miserable backwater" and North America an "anarchic wilderness" if compared to the Aztecs or the Maya. But in the twentieth century, Europe and North America became home to 9 out of the 10 greatest and largest cities in the world.[54]

The Chief Hormone Unveiled

Meet the architects of the shift: Mrs. Efficiency and Mr. Cheap Power. In 1705, Thomas Newcomen's steam engine burned around 45 tons of coal to produce one horsepower unit. Following James Watts' ingenious tweaks, by the late 1800s, the ratio reached 1:1. The efficiency boost of Watt's steam-powered pistons meant that by the 1870s, Britain's 4 million horsepower was replacing the arduous work of 40 million men.[55] Producing more fans than enemies in the British-imitating world (and toxic mercury, smog, and acid rain regardless), with little hesitation the fossil-fueled machine began to reign supreme. And even if Calvinists and Puritans affirmed that there was no Lord but Christ, many found themselves rather at ease with the noble stature of Queen Victoria being dwarfed by the impolite ascent of a new world ruler: the until-then unknown Dark King Coal.

Come his rule, factories grew, and with factories, so did banks and so did cities. But perhaps what a historian like Ferguson doesn't emphasize as he should, is that underlying it all was the petrified, dead organic matter that ultimately powered Britain and its enticing machines.[56] King Coal (and King Coal alone) was the cheeky, smiling Grinch that ultimately sparked industry's unnatural revolution, soon enabling it to spread like wildfire across the globe. By turning fossilized energy into society's new growth hormone, it was it was King Coal, and eventually, his adoptive son Prince Oil, that enabled urban centers in Western Europe, and eventually in North America, to grow curvier and curvier by the hour.[57]

The doped transition into urban puberty was inevitably bittersweet. To be sure, numerous life conditions improved greatly beyond any previous stretch of the imagination: clothes became cheaper, sanitation improved, work became less physically taxing (although more repetitive). But the improvement came with a price tag attached, containing some small-print that was perhaps too unpopular to consider. It's no news that the West built a good chunk of its wealth on the back of slaves and on countless land grabs aided by guns, germs, and steel. But the power boost also took place while taking over the majority of the atmosphere. Home to less than 20 percent of today's global population, today's highly-industrialized nations have come to emit around 70 percent of all the human-made greenhouse gases ever since the Industrial Revolution.[58] And, perhaps unconsciously, and most likely out of enthusiasm or goodwill, a good chunk of it was done in the name of a Father, a Son, and a (Wealthy) Ghost.

Regardless, the discovery and use of fossil fuels begs some questions: If a little charcoal is a blessing, can too much of it be a curse? And is it the case that God provides for our need, but not for our greed?

Fossil Fuels, Individuals, and Individualism

Another key change would come along as cities grew beyond any previous measure: the rise of the individual. Farming villages and settlements had taken around 10 or 12 millennia to establish themselves despite famine, starvation, and pestilence. But in little less than two centuries, coal and oil became the leading underlying forces that accelerated the gradual but steady dissolution of agrarian societies. Already in the 1850s, only around one-fifth of Britain's active population were farmers. By the 1880s, the proportion dropped to one in seven. Come the 1910s, only 1 in 11 worked the fields.[59] Thousands found themselves attracted to the emerging city centers that were growing in fame as hubs of industrial production and international trade.

Once powered by wind and eventually by steam, the use of larger ships also meant that coal-power and luxury items began to replace grass-fed animal power and local goods. Before coal had arrived on the scene, folk felt more or less at home in their villages and settlements. The family, the land, the trade, and the church safeguarded one's sense of being and belonging. But the energetic innovations of the steam engine shook up the agrarian beehive, allowing people to begin to move around much more easily, even as cities and their factories began to grow without end.[60]

This flight to the cities came with a cost, for as early as the 1600s, folk in Europe began to experience a slow but steady weakening of traditional sources of meaning. As nation-states started to grow, a person would be less likely to inherit the family plot or their parents' trade, so his or her place in society could no longer be easily secured as in times past. The son of a milkman could now be forced to end up working for cheap in a nearby spin mill. Thus the shift to coal-power and the emerging urban sprawl began to eat away the traditional path to success. As more and more fled to the cities, the agrarian lifestyle was gradually replaced by a new mode of living without precedents. Sustained to an extent or another by early religious traditions, the sense of being stable and "grounded" in an ecological cradle was soon to give way to economic liberty, to the ability to move around, to competition, and (*tada!*) to the possibility of becoming a footloose individual. "The heir of the city ... is free" would become the slogan of many who left the fields to become people of the city (citi*zens*)—and, eventually, consumers.[61]

Broken at last were the ancient rituals that appealed to the cosmos and that used to stitch people into an enchanted, ordered fabric of cosmic spirits and forces. City life meant freedom for the citizen. Urban centers, and the new religious life that would come to sustain them, began to revolve more and more around the increasing flow of incoming individuals.[62]

However ironically, this urban exodus turned cities into spaces that placed an unprecedented strain on relationships. Uprooted from the land, and now floating in a growing pool of more and more strangers, individuals living in fossil-powered cities would run the risk of becoming more impersonal and isolated. And lacking the nourishment and support one experienced in friendlier agrarian villages (think of the hobbits in the Shire), people began searching somewhere else for direction and meaning.

The Great Disembedding

This uprooting from the land was part of an even greater one, referred to as the "Great Disembedding" by Charles Taylor, a renowned scholar of the secularization of the Western world.[63] The Great Disembedding, he notes, was a set of turning points in civilization when the life of an entire society ceased to have a strong reference to a superior spiritual reality. As villages became more urbanized and controlled by elites, life became more pragmatic: taxes needed to be collected, armies fed, people entertained.

In response to questionable behaviours of the masses and the elites, subgroups of enlightened individuals, such as prophets and monks, began to call kings, priests, and their large social hierarchies into question. And here's the rub: they did so by appealing to a higher, transcendent vision of the human good (as was the case of Jeremiah and other prophets of Israel who appealed to Yahweh and to the laws of Moses). But precisely by confronting kings and their societies *as individuals*, these prophets and monks cracked open the possibility for social life to eventually revolve around the "disembedded" self—the odd person who was brave enough to differentiate him or herself from the whole religious life of larger society. In earlier, less urbanized societies, one related to God (or to the gods) *as a social group* through collective rituals; but come the Great Disembedding, one began to relate to God *primarily as an individual*.[64]

Taste an irony: in the name of reforming the social group according to a higher vision of an "enchanted" reality, *a door was opened for societies to become individualistic and disenchanted*. With prophets and monks pulling the royal rugs from underneath, the marriage of the Great Disembedding with the fossil-fueled growth of cities meant more selfishness and, ultimately, a trend toward godlessness.

Once defined and embraced by land, tradition, family, and collective rituals, over time the "disembedded" individual began to feel more and more lost in the increasingly endless ocean of other disembedded individuals. In turn, what we know nowadays as "existential anxiety" began to creep in. "Who am I?" "Where do I belong?" "Am I worthy?" Being one among thousands in such infinite seas, the answers individuals found to those questions were often wanting. Membership to a clan, a family, a church, or a sect meant less. To fill the vacuum, the new existential gap called for finding one's identity elsewhere.

Taste a second irony: the Great Disembedding was fueled by (a distortion of) the original spirit of Christianity. Following the sixteenth-century Protestant Reformation, European societies gave central importance to the New Testament's call to put allegiances to father and mother and anyone else in second place. The priority was to be given to following God and, in principle, to embracing a heavenly order of *agape*—of self-giving love. Tragically, however, the call was corrupted. Instead of embracing a climate of solidarity that would bring individuals into a new network of brotherhood and sisterhood (as Jesus had called for), people broke existing social bonds but only to begin to organize themselves around personal discipline, political ideologies, inward piety, productivity, etc. The meaning of life was not to be found primarily in serving one another, let alone in one's relationship to the increasingly distant land. Rather, pride of place would fall on one's achievements through one's work and, eventually, on acquiring an increasing amount of purchasable goods (such as candles, which were a luxury item way back then!).[65]

∼

Fossil fuels, growth of cities, individualism, shopping: more or less through shifts like these, the early seeds of what we today call consumerism were sown centuries ago. And all of it to remark on how with the seeds came the weeds, for society's dreams were increasingly choked and prevented from stretching to anything (or Anyone) higher. Factories and fossil fuels made the humanly unattainable city of God look rather boring. The visions of John the seer fell out of fashion. Both the lifestyle of the secular city (and the glimmering gadgets it began to offer) became all-too-attractive to bother looking farther into horizons anywhere beyond.

Further Away, Faster, Cheaper

By living, dying, and eventually decomposing into coal, oil, and gas, God knows how many years it took tons and tons of trees, plants, and animals to absorb countless terawatts of the sun's energy and turn CO_2 into dead underground matter. But we do know that it has taken less than 400 years (and 250 in particular) to undo the job—and the disproportionately rapid undoing calls for a fast-forward, twenty-first-century style.

Back to the 1750s: the time when Britain began to run out of wood and started burning increasing amounts of coal instead. A problem struck: the coal mines were deep and filled with water, so a man by the name of Samuel Newcoming invented the steam engine to pump out the liquid. Then another fellow, James Watt, got creative and improved the engine while turning it into the new horsepower behind manufacturing. Give or take, the Industrial Revolution was born—although likely unaware that King Coal and Prince Oil would soon want to monopolize the show.[66]

Iron rails then made it easier to pull out coal. The rails and the engines paved the way for trains. Then the electric motor came out of the womb in the 1890s once Nikola Tesla discovered alternating current. Then cities began to burn coal to generate electricity. As this happened, Edwin Drake penetrated layers of rock by drilling the first oil well across the Atlantic, in Pennsylvania. Cars started running on petrol. Industry used petrol to create chemicals. Pharmaceuticals prolonged life.

Life became easier.

Populations grew.

To follow suit, German chemist Carl Bosch created fertilizers using fossil fuels and farms began to use oil to power their tractors, which in turn enabled new modes of industrial massive food production.

Food became cheaper.

Populations kept growing.

Populations grew even more.

Cities expanded even as car culture started to become predominant. But World War I broke loose between European nations competing in a fossil-fueled neighborly conflict. After the fight, a depression stepped in, giving way to an ever greater clash. World War II started and ended, leaving behind a monumental industrial complex previously used for producing all sorts of weapons and artifacts of war.

Opportunistic as ever, risk-taking business forces got ingenious and turned the swords into seemingly innocent plowshares. The Industrial Revolution was reborn, but this time running on steroids and soon to wave the flag of full-fledged consumerism. And perhaps nowhere more effectively would marketing contribute to the task, in this case drawing every possible insight from psychoanalysis to create an artificial climate of desire, and to understand (and to manipulate) the human mind.[67]

And manipulate it would.

Seducing Our Minds

Back in 1873, jeans were the denim pants made for miners, cowboys, and prison convicts. But crafty advertisers eventually managed to wrap them around legs of all sorts and sizes: from Richard Nixon's and Marilyn Monroe's, all the way to those of anti-capitalist protesters of the Occupy Movement. Today, more than a century later, jeans have become the most popular garment in all history—even if Levi Strauss is now resting in peace without the slightest idea that his pants would make us all look like miners or convicts.[68] *Bon job, Mr. Strauss!*

Consider, too, fast food. In the early 1900s, the hamburger was scorned in the United States as the "dirty food of the poor." Containing a glamorous mix of butcher leftovers and unwanted ground meat, burgers were cheap food for cheap people. But by spending $1.2 billion in advertising, McDonald's turned them into a seemingly lovely meal, now served to over 69 million people across 118 countries worldwide.[69] Before pumping cheap leftovers into our bodies, the burger giant has stuffed a good dose of illusions into our minds. Smiling Ronald McDonald also capitalized on the brilliant idea of feeding us fantasy mental food while tempting us, willy-nilly, to "Be Happy." *Bon appétit, Mr. McClown!*

Last but not least, cars. Once shared by humans, horses, carts, and trolleys, the streets in the Land of Freedom and elsewhere eventually became monopolized by Big Oil. Car companies lobbied for changing the rules toward greater road support; they helped local police forces to fine and arrest pedestrians who "jaywalked"; they bought up city trolley systems in the 1920s and dismantled them. And the rest is history. As of recently, the advertising bill of the automobile industry spilled over the $31B mark annually. And that's just in the United States. Places like China have increased from less than 10 million vehicles to over 73 million in little more than a decade.[70] Certainly, the freedom of corporate speech has brought the captivity of our collective imagination. Despite the surge, most would still fail to get a hold of the unsacred trick. By featuring (*ahem* . . . using) provocative women in their ads, marketers came to know all too well how to arouse desire. But they came to know even better that such desire was more likely to be appeased by shopping until we're dropping. Regardless, the worship of the car began to spread worldwide. Propelled by cheap oil, the new religion would soon cram the streets, delay commutes, increase the stress, and choke up a good chunk of our daily dose of fresh air. *Mr. Ford . . . bon voyage!*

Call it McBurgers, cars, or denim blue jeans, countless consumable items helped enforce the shopping society, and in doing so, they turned the agrarian lifestyles into an unfashionable trend. Farming became seen as a necessary waste of time for nations now tempted to industrialize without end. Capitalizing on human greed, King Coal and Prince Oil continued to entice both East and West, even if that meant arming themselves to the teeth for a Cold War. Regardless, the myth of progress continued to lure and take root, fully flourishing when the East crumbled and the West soared. Occasionally dressed in Levi's and likely drinking Coke, scientists managed to create atomic bombs, color TVs, and breast implants. Some analysts were so flabbergasted by the bubbly bloom that they would eventually call it "the End of History," rather convinced that Western shopping culture was the ultimate destination of all human societies.

To all this, the sexy hype of the ongoing Hippie Revolution continued to make tradition and religion obsolete. Whether from the East or from the West, ideologies meant conflict. Despite the blue jeans and the burgers, it began to become evident that progress wasn't fully working for the majority of the world, and that traditional Christianity wasn't either. Western societies thirsted for a new story, for a new grand myth. And in doing so they did their best to dream and imagine John Lennon's new world: one with peaceful skies—but with skies alone—making sure to scrape off any leftovers of a sacred heaven above.

Despite the hype, society abhorred the vacuum, and soon the desecrated skies monopolized quickly by the corporate forces of sales and marketing. The time was apparently ripe for lighting up the cosmic screen with the neon lights of advertising. Life was boring, shopping was fun. And so, without hesitation, the powers-that-be became quick experts in captivating people's transcendental desires by turning shopping malls into society's new temples. Thus the gods of old made a comeback, although now dressed up *in vogue*.[71]

But the comeback begged the question: Did religion decline? Or was it the case, instead, that shopping got divinized? Ask the marketing department and, especially, the advertisement industry. Skyrocketing since the 1950s, global advertising expenditures surpassed $642B just around the time of the 2008 financial crisis. And we more-or-less know our history first-hand ever since, now that TVs and smartphones have come to claim pride of place as the most effective means to hook up our imagination and play around with consumers' deepest longings and desires. (*Consumers? Hey! Weren't we citizens just a few pages ago?*)

The Hidden Cost of the Love Affair with Mr. Profit

The hook-up has not come cheaply, or fairly. Even as the world's human population has increased by 220 percent since the 1960s, in that same period, economic consumption has risen by 600 percent, oil use has grown by 800 percent, and natural gas by 1,400 percent. As this happens, the global economy goes on extracting the equivalent of more than 110 Empire State Buildings from the earth's crust every day. In turn, a Princeton ecologist estimates roughly 7 percent of the world's population is responsible for 50 percent of the world's CO_2 emissions, while the three billion at the bottom of the global pyramid are emitting just around 6 percent of GHGs. To all that, being a nation that continues to be secretly governed by King Coal and Prince Oil, well-meaning US citizens have had their free will unwillingly bound to emitting 25 percent of the world's greenhouse gases, even as their ecological footprint is five times the global fair share.[72]

As the religion of consumerism has continued to spread, more recently we have seen even more surges of killer apps and economic innovations. For one, flavored latex products for safe sex are now sold next to candy bars, pop drinks, and beer—certainly a cheap way of seducing us into bed to quench our deepest thirst, placate our existential anxiety, while continuing to numb us to the greater ecological infidelities around us. For another, credit cards have gone viral, offered by the thousands while footnoting all conditions in small print. Fossil-fueled tourism has become mainstream, enticing us to consume all-inclusive destinations that are looking more and more the same. China manufactures cheap exports for all, having emerged as the top burner of fossil fuels and as the world's preferred hub for cheap slave labor. Products are now designed for quick obsolescence largely because of what American technology analyst Neil Postman called a "lust for new technology."[73] The neoliberal myth championed by a small financial priesthood has turned the US into a casino, with a speculative financial sector fueling 40 percent of the country's economy. As democracies are corroded, the world is now ruled by corporate banks living under the religious dogmas imposed by Mr. Profit. And so on—all in the name of a greater and greater slice of the supposedly sacred pie-in-the-sky called "endless economic growth," a pie enjoyed by a greater few while the crumbs are scavenged by the wild rest.

Still, perhaps there is nothing more tragic to all this than realizing how many of us Christians have put on mute the wisdom of the Nazarene, who made it clear that "No one can serve two masters, because either you will hate the one and love the other, or you will be devoted to the one and despise the other. You cannot serve both God and money" (Matt 6:24).

～

Take some, leave some, but such are roughly some of the main foundations and central features of consumerism—or, to be precise, of *fossil-fueled* consumerism, for little of what our cities have become would have been possible without them. The industrial world as we know it today began to be built on the backs of slaves, but a great chunk of the task was eventually transferred to coal, oil, and gas. In fact, if it's the case that "Britain made the modern world,"[74] as it largely did, it's even more fundamentally true that fossil fuels made not only modern Britain and its English-speaking colonies, but every other society that has happily embraced the Industrial and Consumer Revolutions ever since.

Our question at the opening of the chapter resurfaces: Should we take all this for granted? Or could it be that we've fallen into a form of "mass psychosis?"[75] Call it as we may, perhaps twentieth-century British theologian Lesslie Newbigin was right when he affirmed that we don't live in a secular society without God. We live, instead, in a pagan society full of false ones, for the golden calf once worshipped by the ancient Israelites has reappeared. King Coal and Prince Oil have indeed struck once and back again, continuing to do their best to push their dark fossil force right through our veins.

So who knows what Jeremiah or John the seer would have said about this energetic love affair of ours? But maybe former Oxford economist E. F. Schumacher was not exaggerating by remarking that the modern economy is propelled by a "frenzy of greed and indulges in an orgy of envy," which, in turn, he saw not as accidental features, but as "the very causes of its expansionist success."[76]

Babel Remixed

If images say more than a thousand words, what to say of the front cover of *Time Magazine*'s 2006 "Person of the Year" edition? The picture highlighted an iMac, followed by a small legend below:

> Yes, you.
> You control the Information Age.
> Welcome to your world.

The picture was even more provocative considering it had a foil-like, reflective film attached to the iMac's screen—one could see a distorted picture of one's self reflected in it!

Next to one-click shopping, our digital era has brought easy alternatives to socializing. Snapchats, Whatsapp, Facetime and more have become the top hits of the day. Distances apparently become short as time collapses into what has been called a "perpetual present," a situation in which we have lost the capacity for remembering the past and for dealing with time.[77] All Visa cares about is that we "Go." It's easy for us today to immerse ourselves into the virtual overflow and in doing so run the risk of devolving into increasingly cybernetic beings. Paraphrasing the apostle Paul, it's tempting to "change the glory of the real human being, for images and appearances of the virtual human being" (Rom 1:23). But despite their worth and charm, we may be paying the price by interacting with more and more images, screens, and virtual appearances.

What if one of the most influential literary works of Western culture, the Book of Genesis, is allowed to speak afresh into our ecological crisis, even as it addresses our virtual condition in doing so? And what if we approached Genesis not as a book of modern science or dry, factual journalism, but as a literary masterpiece full of wisdom, irony, and drama? I sense that German scholar Gerhard von Rad was right in recognizing that the book is better characterized as "sacred knowledge" narrating a subversive story of origins.[78] Again, Genesis is not so much interested in describing what exact sequence of physical events brought us here (a modern scientific question). Instead, the book engages with more fundamental realities: "Who are we?" "In whose world do we live?" "Why are we here, and who's in charge?"[79]

In chapter 4 we'll revisit these questions, but here we put our analytical lenses to the side to immerse ourselves into a contemporary retelling of one of the turning points of the holy book: the ancient story of the tower of Babel.[80]

Bricks, Bricks, and More Bricks

Twice does the narrator of Genesis tell us that the living source endowed humanity and all other creatures with the gift of reproduction and the task of filling the earth (1:28; 9:1). In turn, we are told, the Eternal One blessed humans—the *soil*beings—with the possibility of multiplication, summoning them to extend a sacred garden toward the ends of the earth. However mysteriously, Yahweh endowed humans with the risky task of co-creating and unpacking the underlying potential of the cosmos. They (we) were to turn a world of sounds into symphonies, of sand into sand castles, of fruits and legumes into flavorful banquets.

With simple literary brushstrokes, the story of Babel paints a sketch of the divine response when the opposite happens. Babel is the sacred garden, gone wrong.

At odds with Yahweh's intentions, the peoples of Babel thought rather highly of themselves. Too quickly, and without consultation, they bypassed the divine call. They crafted their own plans, with their own power, while saying to themselves: "Let's make bricks and cook them!" And then they said: "Come, let us build ourselves a city, with a tower that reaches to the heavens, so that we may make a name for ourselves; otherwise we will be scattered over the face of the whole earth" (11:4).

"Let us . . . ourselves . . . let us . . . ourselves." No reference to Yahweh, no reference to the living world. Babel was a gong show of autonomy and control: human plans, with human power, to transcend all things human. The first bite of the juicy fruit was not quite enough. As descendants of Nimrod, the city builder and violent hunter, the people of Babel wanted to drink from the heavenly chalices in the company of the gods. Confident in the power of their enticing technologies, they built a hefty imperial city with a royal temple at its core. Babel worked endlessly to create a stairway to heaven; Babel sought to bring the heavens down, always longing for a royal sip of the divine. Walking away from the morning bounty of Eden, Babel lit its courts with neon lights of self-sufficiency. And they did it all in Yahweh's face.

But their production lines for the baking of their bricks did not prove to be quite enough. Searching for a trademark to set themselves apart, the people of Babel crafted a name unto themselves, making sure to exclude the Living One at all costs—even from the small print. The external threat of being scattered was too hard to swallow. They tried instead to consolidate operations to avoid a price crash of their stock. Babel couldn't afford to lose a single slice of the market: one city, one temple, one language, one name. Monotony and hegemony; fat wallets, tight fists. Claiming to be as great as gods, the peoples of Babel dressed themselves up as best they could, with lips shinning in glimmering lipstick. (*Muak!*)

To all this, quite surely, there was room for satire of the sacred kind. The CEOs of Babel wanted to bring the heavens down, but even their best efforts fell short, for the One Above was forced to step down to be able to see the city and their tower. Nothing ultimately impressive. In fact, rather disquieting: Seeing the danger and the boredom of their mass production schemes (bricks, bricks, and more bricks), Yahweh saw through the make-up and the glitter. Engineers and shareholders growing rich; cheap laborers drying off the sweat from their brows.

Short story shorter, Babel became the archetype of an entire society systematically organized against the purposes of Yahweh. If the early humans rebelled individually, Babel rebelled as a collective. Instead of spreading elsewhere to exercise dominion, Babel unified around one voice to have domination, becoming a primitive prototype of social sin.

Empire, Then and Now

To all that, the Almighty introduced a virus to the system—confusing their coded language to interrupt their operations. But the viral confusion was not enough: the language changed, but their hearts didn't. After being scattered throughout the earth, the spirit of Babel kept finding reincarnations. Centuries later, after Egypt and Assyria, came Babylon in imperial splendor, and with Babylon came the same old story: Babylon was the agricultural, religious, and political superpower of the sixth century BCE. In Babylon, one found splendid temples, the loftiest cultural artifacts, the height of power. Everything shone in Babylon.

However, as with Babel so with Babylon: the empire rose and the empire fell. Nowadays, the hype is elsewhere: not in Babel or in Babylonia, but in the virtual world of California, now that Apple (and Facebook and Amazon and Google) are taking over the cloud in a new corporate attempt to unify forces, and, in one voice, bring the heavens down. Apparently, the forbidden fruit is not satisfied with flooding the world with iMacs and iPhones. Apple wants more. In fact, it wants it all. Gigabytes and gigabytes for each of us to fill up cyberspace at the beat of our whim: songs, pics, clips, apps, and endless snaptricks. One can almost upload one's life. The tech-giant wants it all gathered up, wireless, in a single cloud. One world, one corporation, one operating system, one logo (bitten and remastered). *"We've got the whooole world in our hands."*

The concentration of power is little more than the reappearance of Babel, but this time remixed into a great virtual cloud gathering around the altars of entertainment and confusion.

The same, of course, must be said of the power-grabbing attempts of the oil and gas corporations. The astronomical salaries of their executive priesthood make it look as if the living world actually belongs to them. What is not so evident is that the track record of King Coal and Prince Oil is nothing short of wreckage to the global climate: they are Babel grown large, running on fossilized steroids.

∼

You'll excuse the satire and sarcasm. I really am an admirer of the tech world and its gadgets. I do enjoy controlling my Bose sound system through my cell phone and using my Whatsapp to share instant pictures with my loved ones. Fossil fuels and digital technology have brought us endless wonders, and for them we can be thankful. But it's not my intention here to repeat what we already know about the glittering powers of the internet and our mobile phones—both of which, by the way, have become the world's sixth largest consumer of (mostly fossil-fueled) electricity.

All this is to note that if we ignore our history we're prone to repeat it, magnifying its consequences. The imperialism of Babel is no novelty: empire has been with us for a good 4,000 years or so, and it has hardly been the exception, but the rule. Concerted attempts to control and impose a social order have been the flavor of the day for ages. After Babylon came the Persians, then the Greeks, the Romans, the Byzantines. In their day, all of them giants; but today they're relics of old. Their might and glory were either covered in ashes or now overshadowed by the U.$. of A. (or by China, India, Russia, or Brazil who are now lining up right behind).

Wisdom from the Past?

The towering architects of Babel leave us with at least three lessons for our day and age. To begin with, the story of Babel is the first of many in the Hebrew Scriptures that show the clash between the gracious gift of Yahweh and concerted efforts of human pride and autonomy. Babelian-type operations are a confrontation with the living source, and sooner or later they lead to a heavenly counter-operation aimed at confusing those determined to resist the grain of the universe. The story of Babel is a divine act of opposition; a firm "no" to the human impulse to do away with limits or live without reference to the divine.

Still, Babel is also an act of fresh grace: grace in the face of imperial regimes, grace to those who are controlled by such regimes, grace of lifting away the burden of anxiety that comes by spending our energies making a name unto ourselves. By confusing Babel, the Most High liberates humanity from the restless effort of trying to be the gods we're not.

In a day when the same banks, coffee shops, and burger restaurants replicate themselves endlessly in every corner (in Chicago, Bogotá, Paris, or Rabat), the story of Babel is one of hope. It pulls us into a drama in twelve dimensions that is far more interesting than those of Hollywood or of Wall Street. For those who have eyes to see and ears to hear, the story of Babel reveals the Living One resisting cookie-cutter techniques and mass production schemes: "One language, one city, one tower, one name." Babel is an act of grace, recalling the living source's preference for rainbows of multicolored cultures, instead of a monotone virtual cloud filled with increasingly lonely people who are either doped or deaf to the state of our only planet.

Against boredom and monotony, against repetition and uniformity, the scattering of Babel recalls how Yahweh knitted us uniquely in our mother's womb (Ps 139). The story of Babel is about opposition and grace because Yahweh wants unity among humans; but not unity around virtual purposes or twisted schemes.

In our warming, hyper-digitalized world, the outcomes of Babel warn us about what happens when we sacrifice diversity in the human community and the living world. The narrator of Genesis reminds us that homogeneity created by cultural and technological control is an affront to the Most High. So is the human aspiration to bring down the heavens that goes with it. Babel shouts aloud, always saying "no" to the Creator but "yes" to the seemingly unquenchable lust of our self-serving plans. According to Jeremiah, Babel is like cheating to God while sleeping around with the fashionable deities of the day.

A Contrasting Vision

To all this, the Book of Genesis contrasts the story of Babel with that of Abram, a descendant of Shem. ("Shem" stands for "name" in ancient Hebrew.) Abram, descendant of Shem, is the sacred answer to the autonomous desire of the giants of Babel to make a "name" for themselves. The descendants of Shem were called to rewrite the story of Babel, in reverse. Thus we're told that the living source encouraged Abram to leave his country and his family by giving him a promise: "*I* will make of you a great nation, and *I* will bless you and *I* will make your name great so that you are a blessing . . ." (Gen 12:2).

"I will . . ., I will . . ., I will . . ." "I, I, I." With Babel left behind, the narrator tells of Yahweh the living source taking center stage, now showing favour to Abram and his offspring, through whom all the other nations would be blessed. Amidst the confusion of languages and the flavorsome entertainments, the Most High set an eye on a nobody and his wife, struck as they were by 75 long years of age. An elderly couple willing to follow an absurd promise was enough for the creator of impossibilities to reverse Babel's trend—and to bring life into Sarai's worn-out womb, whose ovaries were all too wrinkled for any good. Through their offspring, the story goes, Yahweh would bless all the peoples of the earth.

Despite having been called to be a light to those around them, however, Abram's family soon became a recycled copy of its surrounding empires.

From day one, the Israelites were great admirers of their neighbors. They wanted to be like Egypt to the south or like Assyria to the north. They envied the spotlight, building high altars like those of Babel, and living extravagantly as they did in Babylon. (Who wouldn't?)

∼

Are we also fascinated by the spirit of Babel today? Does that spirit still whisper in our ears, with words too subtle, and perhaps too tempting? Our ecological track record outlined above proves that we love to build our cities and corporate towers to bring down the heavens, but that we've done so without much reference to the cosmos nor to the living source that sustains it. Is the One who descended from heaven all too boring and *passé*? Has Adidas put Abraham's dusty sandals out of fashion? It surely seems more entertaining to amuse ourselves with the latest apps of the iPhone X. We'd rather be embraced by the virtual pillar of cloud that moved away from Babel and Babylonia to now settle in California. The cloud promises to relieve our anxieties and hide our imperfections. The cloud enables us to swipe left the wounds of the world around us.

But there might be more to reality than meets the eye, for the story of Babel is not merely a great "no." Babel's fruit was taken away from us because to live as they lived in Babel is to shy away from the great "YES," "YES" to eating gratefully from the thousand and one trees whose fruits were freely given to us all, "YES" to a wonderful world that's Yahweh's work of gracious love.

How can the Christian gospel be a source of good news in today's world, living as we are in a day and age of looming ecological collapse?

4

──Climate Change & the Good News──
A Glimpse into the Ecological Vision of the Scriptures

"Why should we expect prayer for mercy to be heard by What is above us
when we show no mercy to what is under us?"

—Pierre Troubetzkoy, American painter

"God has a dream."

—Desmond Tutu

THE PRETENSIONS OF BABEL and the tears of Jeremiah did not have the last word. Even as swirls of grief stirred the prophet after seeing how his contemporaries embraced the ways of Babylon's empire, Jeremiah saw beyond. Against the odds, Jeremiah foresaw an age when the living source of the cosmos would write a new song in people's deepest self. He glimpsed a day of healed wounds and health restored; a day of cities rebuilt and of bitter mourning turned into brooks of water and radiant joy. He also had hints of a time when an outlandish servant king would become the chief of the nations, bringing forth an unimaginable time of renewal.[81]

Centuries later, when a traveling Galilean peasant entered Jerusalem in tears in the last week before his execution, this promise was writ large on the horizon of his aspirations. Like most common folk of his day, being the peasant that he was, he also longed for the day when Yahweh would liberate the land from the shackles of military occupation, from unjust taxation, and from the glittering but oppressive forces of evil now embodied in the Roman empire. Rome and the powers Rome embodied had to go, and so did the power-hungry religious elites that controlled and corrupted Jerusalem. Like most sidelined Galileans of his day, the peasant longed for a Mount Zion of peace, of fairness for all; a time when those who hungered for righteousness would be satisfied and where the peacemakers would be called children of God. The peasant knew that Yahweh's actions in history would lead to nothing short of the renewal of all things: to the great come-down of heaven to earth.[82]

The come-down he announced, however, certainly did not take root all at once. Enough to open our windows or turn the pages of a newspaper to see that the world is still wounded by injustice today as it was then. And clearly, the renewal did not take place in a way that anyone expected, then or now. The itinerant Galilean did not defeat the Roman prefect who sentenced him to death, nor did he rapture his followers by taking them on a magic trip into the vast blue skies. In many ways, the climate remained the same after his execution. The high priests and rich aristocrats in Judea kept putting burdens on the backs of the peasant poor; Rome's soldiers kept taking others down; families and villages continued to suffer the economic pressures of the Jerusalem Temple, of taxes to Herodian kings, and of tributes to Roman conquerors. For all they knew, the dream of a new liberation promised by the prophets was crushed by Rome's might. In more ways than one, the leader behind the Galilean protest movement seemed to be a preacher of yet another failed peasant dream.

Unless, of course, the good news (the gospel) of the great come-down was altogether different—far more subtle, and, perhaps, far more dangerous. "Gospel" meant *euangelion* in the ancient world; it meant the announcement of victory, the proclamation of good news about this or that ruler coming into power. For Isaiah the Israelite prophet, the good news had to do with proclaiming the reign of Yahweh over all the living world; for Saul of Tarsus, the gospel was about declaring that the heir to the throne of David became the lord of the cosmos and as the head of all rulers and all powers. In fact, Saul of Tarsus went on to affirm that the regeneration that came in and through the Galilean's messianic mission happened to be nothing short of the recreation of all things.[83]

Over the course of its long history, however, the church has either trimmed, distorted, or misunderstood the vast scope and the earthiness of that accomplishment. Today, we've inherited a vision that Wendell Berry rightly denounced for its restless anxiety about "incanting anemic souls into heaven,"[84] while ignoring the social and ecological ripple effects of the new beginning.

Why have we gotten here?

God, Ceasar, & the Shopping Mall

Recall again the fish who never knew what it felt like to be wet because being wet was all it ever knew. Like fish born into stagnant waters, today we take box stores and malls for granted, often ignoring their murky economic history sketched in chapter 3. Being raised in our shopping societies makes us blind to what shopping actually feels like. Worse, as the religion of consumerism has mutated over time, it has gradually seduced and taken over our imaginations, leading believers and skeptics alike to forget or ignore the Hebrew Bible's almost obsessive emphasis on the land.[85]

The history of why this happened is long and complex, and so are its repercussions. Here it's simply worth noting how some Christians today often live as if heaven and Earth are two separate realms, and as if discipleship is about fleeing from the things below to reach those above. Theologians refer to this split as "dualism" or "escapism." Others (among whom I count myself) have been led to believe that the good news are centered on humans; that the gospel is about something that happens *to us*. Intellectuals call this "speciesism" or "human-centeredness," which more or less shakes hands with some ancient Greeks who assured that we humans are the crown and the high point of all things. Common as well is to maintain a rather narrow view of Christianity, convinced that the faith is pretty much about turning away from one's personal sins, believing a theological formula to secure a place in heaven, and then endlessly convincing others to do the same until the world comes to a supposed end. This turn-or-burn religion has been popularized by "The Five Steps to Salvation," or by top-selling novels about the unfaithful being "left behind" even as the saints are (supposedly) raptured into an immaterial heaven. And yet still others amongst us have been taught to live as if Christianity and politics were like water and oil, affirming that religion is a matter of personal choice unrelated to politics and economics, which are seen as dirty matters of this world—a sort of yin-yang gone public.[86]

Generation after generation, we've inherited a history marked by mutations like these, often watering down and distorting the original intention of the new beginning. In doing so, however, some people of faith often ignore that the entire biblical tradition is about restoring the right relationship between Yahweh, us humans, and the rest of the living world. "Blessed are the meek, for they will inherit *the earth*" (Matt 5:5). Today God is popularly perceived as a being who is either too distant, too domesticated, or—ironically—too small (a trend reflected in books by new atheists, such as Christopher Hitchens' *God Is Not Great*). On top of that, the inward-looking temptations that elevate the spiritual above the physical have led many Protestants and Catholics to believe that issues like climate change are of lesser importance when it comes to working out one's faith.

But by denying God, or by watering down the gospel (the *euangelion*), two great doors have been left open. For one, beastly corporate powers have taken over. Pharaoh, Babylon, and Caesar's empire have all indeed struck back. And the odds are great, for as Wendell Berry observed, "Caesar is no longer a mere destroyer of armies, cities, and nations ... he is a contradictor of the fundamental miracle of life." For another, faith has often been reduced to a largely individualistic experience, consumed as if it were aspirin for personal comfort, but stripped of its economic or ecological ripple effects.[87] In a day and age when attending some church services is sometimes similar to visiting a spiritual shopping mall, perhaps one of my mentors, Rikk Watts, was right in saying that in the West we have turned God into a "comfortable slipper."

Church & State: Separated or Isolated?

Consider a second dilemma we face today: faith in God is either used conveniently by some politicians to sanctify their power (very much like Jerusalem's chief priests did in Jeremiah's day), or marginalized at all costs. In today's world, anyone can publicly appeal to Adam Smith, Thomas Jefferson, or Albert Einstein, but religious discourse is increasingly sidelined from public life. Upholding Steve Jobs as an example of leadership is seen as being cutting-edge; quoting Moses or the apostle Paul in a boardroom could be a recipe for getting fired. Religion is (religiously!) believed to be a personal decision left to each person's choosing; science, the laws of economics, and rationality continue to be seen as the supposedly objective ways of directing a nation or a company.

Part of this split between public facts and private feelings has come about for understandable reasons. Religious beliefs *have* been abused in public for wrong purposes. Christianity meant crusades in the fifteenth and sixteenth centuries; it meant thirty years of genocide between Protestants and Catholics in the seventeenth. As expected, the West has longed for a more peaceful story ever since—and, until lately, the Age of Lights has risen to the occasion.

Partially in reaction to all the blood spilled in Christianized societies in the very name of Christianity, the light of science and reason was perceived as being able to pave a new way. Wanting to counter some of their own religious horrors, societies in Western Europe and North America gradually squeezed the beliefs of religion *into* one's heart, even if that would eventually mean casting them *out* of public life. In the eighteenth and nineteenth centuries, God was kicked upstairs and the living world and society began to be seen as a machine that could be studied and operated by mastering the neutral laws of nature. In short, St. Augustine and St. Francis were out; Copernicus, Newton, and Darwin were in.[88]

A paradox came along with the shift, as public affairs became increasingly insulated from religious critique. Freeing the state and the market *of religion* eventually led to the belief that politics and economics were free *from* religion. And instead of allowing religious talk to take place within the *separation of* church and state, many societies in the West ended up with the *isolation of* the church *from* the state, and vice versa. Ever since, a famous statement has been subtly (and ironically) misinterpreted to fit the modern way: "Religion" belongs to "God," the market and the government belong to "Caesar."[89]

Since then, most of us have been taught that in public, one is liberal or conservative or green or socialist, and that in private, one is Jewish or Buddhist or Christian or atheist, very much as if one's religion was a matter of choosing candy bars in a supermarket. But regardless of one's convictions, pushing Christianity to the side has ended up by paving the way for the state and the market to become the new towers of Babel. Today, national governments and corporations have become modern gods—more or less contemporary versions of the old sea-mammoth Leviathan once depicted in the Book of Job (3:8). We have opened up a glittering Pandora's Box in the name of consumerism and secularization, and by doing so, in the words of William Morris, we have been "cast into the jaws of this ravening monster, the World-Market."[90]

A Climate of Desire

Yahweh's Earthly Temple

The people who wrote the Hebrew Bible engaged with the world and experienced its living source quite differently than the majority of us who live amidst the concrete (and now virtual) jungles of our industrialized nations. And it's by remembering their history that we can more fully grasp the ecological ramifications of the good news for the world remembered in the New Testament.

As with Jeremiah, the Hebrews had other prophets calling them to embrace an alternative version of reality. Being the peasants that they were, these folk remembered their earthly origins and foresaw an age when Yahweh would act again in history, in a new way. "And it will come to pass afterward, that I will pour out my Spirit on all people," announced the prophet Joel somewhere between the ninth and fifth centuries BCE. Likewise, ancient priest Ezekiel, son of Buzi, received visions that cut through the smokescreens of Babylon's as the people of Abraham remained in captivity in foreign land. The visions stretched into a future where the people would be restored to right relationship with Yahweh, with each other, and with the land. "This is what the Sovereign Lord says to these bones: I will make breath enter you, and you will come to life . . . I will put my Spirit in you and you will live, and I will settle you in your own land" (Ezek 37:5, 14).

Burdened by the anxiety of being displaced into a foreign empire, the prophets nevertheless reclaimed the past, the origins, the source—the time of the sacred garden where the soilbeings were formed out of dust and breath: "Yahweh formed *adam* out of the dust of the ground [out of the *adamah,* in Hebrew] and breathed into his nostrils the breath of life, and *adam* became a living being" (Gen 2:7).

Adam, adamah; bones, breath, Spirit, land—all engaged in a sacred dance, according to the poetry of these ancient writers. Theirs were visions of hope. In new creative fashion, Yahweh would recreate people again, in a new garden. In the time of renewal, Eden would be made new, and a river of fresh water would flow into the sea. And wherever the river would go, said Ezekiel, every living creature that swarms would live. Even the waters of the sea become fresh; everything lives where the river goes. And on the banks, on both sides of the river, there will grow all kinds of trees for food. Their leaves will not wither, nor their fruit fail, but they will bear fresh fruit every month because the water for them flows from the sanctuary. Their fruit will be for food and their leaves for healing.[91]

Bursting all bounds of rational talk, around the years of his exile in Babylon in 593–571 BCE, Ezekiel spoke of a day when the living source would come down to heal and restore. The time would come when Eden's river of old would flow from the sanctuary, from the holy place. Ezekiel saw a new garden, a new city, a new temple; but he envisioned a different temple from Babylon's temple, he spoke of another reality different from Babylon's version of reality.

Temples Made by Human Hands

Shrines and temples popped-up all over the ancient world. Peoples of old saw them as the sacred place where heaven and earth merged in one, providing a dwelling place on Earth for the gods of heaven.

Babylon's was first in line. Erected by slaves in Sumer's capital city, their temple honored Marduk, revered by the Babylonians as the son of the chief god of the heavens. Seeing him as the supreme guardian of life, Babylon's founding myths tell of the epic battle of Marduk defeating the seawater goddess Tiamat. The story goes he shot an arrow in her throat and went on to create heaven and earth, the stars, and other living things out of her corpse. Said briefly, for the Babylonians, creation came about through war, and Marduk's supreme kingship through military victory, a sort of a primeval version of Game of Thrones.[92]

Taste the opium of the Babylonians and their mythological marketing plan, which was rather clever: Given that Marduk is a violent god, Marduk's representative on earth (the Babylonian king) had the license to behave likewise. And given that Marduk reduced Tiamat's companion Kinguu into servitude to then kill him and eventually mix his blood with earth to create humans (who were, in turn, to perform forced labor for the gods), so could the Babylonian king. The king of Babylon could have slaves providing for him because, for all they knew, he was the representative—the image—of Marduk, whose might was supposedly writ large in heaven.

Mythical stories like Babylon's came alive in temples built by human hands. Home to a massive statue—to an image—of the god in their midst, life in ancient cities revolved around these monumental stone buildings. Like today's skyscrapers, ancient temples gave cities fame and prestige. Temples shone, and their artwork served as visual architectural propaganda to bring into life the ancient myths concerning the origins and meaning of the cosmos. Lacking TVs and iPhones, temples were the mass communication media of ancient times, filled as they were with images and carvings displaying out loud the subordination of slaves and proclaiming the power of human kings—kings who were hailed as the earthly representatives of the temple's god.[93]

The Long-Awaited Marriage Feast

Ezekiel and other biblical writers said "yes, but not quite"—so close, but so far. There were no stone temples in Eden. The story of the Bible begins and ends without any sanctuaries made by human hands. And although the Israelites did come to have a stone temple built by King Solomon, the first-century writer of the Letter to the Hebrews was quick to note how that building was very much like a sign on the road: useful as a pointer, but not needed anymore once travelers reached their destination.[94]

Unlike Marduk's temple in Sumer, the Book of Genesis tells of a garden without statues of stone. We hear instead of a paradise inhabited by rather different statues: ones of flesh, hand-sculpted not out of dirt and cold blood, but out ouf dust and sacred breath. The poetic sketches are subtle but clear: animated by Yahweh's living breath, human beings are the living statues placed in a living garden that is nothing short of Yahweh's earthly temple. That's why in revolutionary fashion Moses forebade the Israelites to make images of God from wood, metal, or stone: *because we humans have been entrusted that role.* Genesis's sacred wisdom cracks open the story of life, a living drama to be carried forward by all of Eden's kingly and queenly priests. The biblical story of origins begins in a living temple with living beings called to be in living fellowship to represent the living God.[95]

But the story doesn't end in Eden. Centuries later, in the final scenes of the Book of Revelation, John the seer portrayed a liberated community of life—a place also without a temple building. Being the poet that he was (and not a crystal-ball future-teller), in candid metaphor John portrayed a radiant garden-city descending from heaven to earth. The city had streets of gold and a pacified crystal sea; it was a place of harmony between humans and God, between culture and nature. John's images spoke of paradise set free from thorns and thistles; he saw Eden come of age, infused into a vast city spiraling down in an eternal dance of light and cosmic peace.[96]

The days of Babel's sweatshops and the endless working hours in Egypt were gone. Indwelt by the breath, the seer's final brushstrokes portrayed the wedding of eternity and history. After witnessing the fall of Babylon and her adulteries, John went on to see the radiant city dressed as the bride of the Lamb, and he glimpsed the long-awaited marriage feast as he peeked through the curtains. John's liberated imagination foresaw a place without a temple *because he saw the entire cosmos* becoming the hallowed space where the overflowing presence of God saturates everything and everyone. There, at last, was Yahweh the living source descending in fullness to dwell among humans in the banquet of joy of the eternal Sabbath. There, at last, was the community of life embraced by the peace she had always desired. There, at last, were heaven and earth united forever in cosmic ecstasy and sacred intercourse without end.

A songwriter of old had had a glimpse of that hope:

> *Praise Yahweh, sun and moon,*
> *praise Yahweh, all you shining stars!...*
>
> *Praise Yahweh from the earth,*
> *you great sea creatures and all deeps,*
> *fire and hail, snow and mist,*
> *stormy wind fulfilling his word!*
>
> *Mountains and all hills,*
>   *fruit trees and all cedars!*
> *Beasts and all livestock,*
>   *creeping things and flying birds!*
>
> *Kings of the earth and all peoples,*
>   *princes and all rulers of the earth!*
> *Young men and maidens together,*
>   *old men and children!* (Ps 148:3, 7-12, ESV)

The rest of the biblical authors unfolded the story of what has taken place between Eden and the garden-city; the story of Yahweh patiently but steadily overcoming the cosmic forces of darkness that have muddled the show. The biblical narratives contain the sacred memories of Yahweh confusing Babel and confronting Pharaoh, king of Egypt. They recall the living source of creation rescuing the ancient Israelites out of empire, calling them to display an alternative lifestyle. Israel was meant to be a society of rest for all every seventh day, of respect for the land and its creatures, of distributed property and forgiveness of debts, of care for the widow, the orphan, the foreigner. Israel meant engaging with Yahweh to sow and cultivate a culture of love.[97]

A Climate of Desire

Mistaking the Shadow for the Light

Despite the rescue out of the empire in Egypt, and despite the lofty calling they were given, the Israelites came to mimic the ways of other nations, giving way to the dark agents of evil disguised as light. Like their earliest ancestors, they were also seduced by the poison hidden in the alluring voice of a crafty creature—the ancient serpent symbolizing the forces of autonomy and domination embodied in all empires, old and new. Since early on, the Israelite leaders found themselves wanting towers like Babel's and temples like Babylon's.[98]

Eventually, Moses gave them a traveling tent in the desert, and then Solomon a relatively sober stone building in the promised land. Two shrines, each of them seen as the resting place for the *Shekinah*—the presence of Yahweh dwelling on earth. But neither tent nor temple had stone statues. There was only a veil with embroidered angelic figures safeguarding the symbolic resting place for the sacred presence of Yahweh.

Still, the Israelites confused the road-sign with the destination. They took pride in themselves as if they were the sole custodians of the living God. They overlooked the notice that Yahweh did not live in temples made by human hands, but that heaven was Yahweh's throne, and the earth was Yahweh's footstool.

"The temple of the Lord, the temple of the Lord, the temple of the Lord!" they eventually exclaimed, trusting in themselves, very much as if they could forsake justice and peace while turning Yahweh into a comfortable slipper and a tribal deity at their command.

"Not so fast," replied Jeremiah. "You've turned the temple into a den of thieves. Yahweh will abandon it, outraged . . . even if heart-broken and scandalized in tears . . ."

Said and done. It was a matter of time before the Babylonians tore down the sacred building when their king Nebuchadnezzar II besieged Jerusalem in 587 BCE. And so the old certainties of their uniqueness as the chosen people of Yahweh were suddenly put in jeopardy. Their political lust and their misplaced desires put the descendants of the Hebrews in exile.[99]

An Alternative Temple Movement

Centuries passed. Empires came and went. The exiled ones returned to the land. And as history recycled itself, eventually king Herod—called the

Great—and his fellow aristocrats rebuilt Solomon's temple, supposedly inspired by the bright day that the prophets had foreseen after the long night of exile. Jeremiah and Ezekiel had in fact spoken of the presence of Yahweh returning to dwell in their midst. Beyond the doom of empire, the poets envisioned a time of restoration of their fortunes and of the rebuilding of their temple. But no one ever quite foresaw how—especially seeing that Pontius Pilate and almighty Rome had taken over.

It was then that the outcast came. Conceived out of wedlock and turned into a refugee as a child, he always stood on the edge of the inside. Native from the poor northern province of Galilee, he grew into a traveling prophet, eating with outsiders, feasting with prostitutes and ostracized tax collectors, healing people on the sacred day of the Jews. The outcast was like a magnet, attracting many, repeling others.

And in doing so, he lived against the odds. Not having a place to lay his head, the itinerant preacher was meek but wise, compassionate but brave. So brave that his followers remembered him for claiming to have authority over Herod's temple, for releasing folk from unclean spirits and their sins, for delivering people from the powers of darkness wherever he went, for feasting regularly with rich and poor, both with the healthy and with the sick. The outcast stood for all that the temple had stood for, even if he acted without the permission of religious leaders or high priests. At one point, he even entered the supposedly sacred building to overturn the merchants' tables and lament that Yahweh's international house of prayer had become a sectarian den of bandits.[100]

"Destroy this temple," he cried out in a loud voice, "and in three days I will raise it up!"

The itinerant Galilean, it turns out, had gone around proclaiming that the great come-down of heaven had arrived. Having started his mission in the wilderness in the presence of wild animals, he eventually announced that the days of Jerusalem's stone building were over. Regeneration and forgiveness were no longer restricted to one place. As John would put it, Yahweh was now going live by pitching down a traveling tent to show up in the neighborhood, human style. With Cosmic Consciousness becoming tangible flesh, heaven's living groundswell of renewal began here on earth. And so the trees began to clap their hands and the mountains sang for joy because the temple's candlesticks were no longer needed because the sun's lips were already kissing the horizon. Creation at large was finally beginning to experience the climate it had always longed for and desired.[101]

A Climate of Desire

The Rejected Stone, the Cornerstone

The outcast's native name was Yeshua bar Yosef—although he is best known to us today as Jesus son of Joseph. And the temple he spoke about was not a temple made by human hands. As far as his earliest followers testified, the temple was his very self: Jesus's earthly body was the very embodiment of the presence of Yahweh (John 1:14): he was the true visible living statue of the true invisible living God (Col 1:15).

Even if the elusive Nazarene had more to him than met the eye, Herod and Pilate and those in power could not see past the veil beyond which Yahweh was hiding. All they saw was a preacher in sandals followed by the people; a would-be Messiah who had, in fact, become a threat to their power. That's why like most tyrants, past and present, the *políticos* decided to do away with him, mocking him as a false, populist king.[102]

In a sense, they were not far off. For all they knew, it was not from London, Paris, or Rome, but from Galilee that Jesus came from. In the outcast, they saw little else than a failure. The authorities believed that true kings looked like Herod or Caesar. There were convinced that the gods sided with Rome, not with Nazareth.

The shrines of Rome's empire throughout the Mediterranean had sculpted images of strong, deified Roman emperors subjugating weak nations under their might. In one, Claudius Caesar dominated the female personification of Britannia; in another, Nero Caesar the female representation of Armenia. Both Britannia and Armenia were mostly undressed, and so were the muscular emperors who pressed down their bodies, subjecting them as their new concubines. (Sexual) imperial theology at its best.[103]

Still, John the seer eventually went on to proclaim that the Nazarene was "ruler of all the kings of the earth," (Rev 1:5) an absurd proclamation then as it is now because Jesus was the sort of king who washed the dirt off the toenails of his disciples and who sat down for dinner with folk of unglamorous reputation. And absurd too because did not enter the holy city on a white chariot holding a MasterCard in one hand and a lightsaber in the other. He entered Jerusalem in tears, on a donkey, wearing his heart on his sleeve, ready to surrender himself to the darkest fate of the dispossessed by dying naked on a Roman cross.

Crosses were a scandal back then. They were the most ferocious weapon of the empire reserved for political insurrectionists and the scumbags of society. Crosses were the equivalent of nineteenth-century electric chairs, used to silence and bring public shame on dissenters and troublemakers.

They were the ultimate instruments of injustice that allowed the politics of Rome to try to crush down the politics of Nazareth.[104]

Conscious of the cost, the Nazarene decided to stride through the valley of death nonetheless. Praying for his enemies, Jesus went all in, making his the destiny of the victims of evil while suffering the punishment deserved for all evildoers. Well aware that Rome crucified dissenters by the thousands, Jesus nevertheless surrendered himself to the ultimate political and military weapon of the enemy. He allowed Rome to do the very worst to him. Willingly, he suffered unjustly to expose the shame. The violence of the cross would be evil's best shot to silence his revolution, but deep down he knew that the twelve inches of steel forged into Roman nails would eventually rip the heavens open. Jesus knew that the power of Yahweh's love was greater; he knew that the cross would finally unmask and destroy the ongoing grip of the violence of this world, which so often serves the wicked but harms the innocent.

Thus in him some saw a moron, a reason for shame, a would-be image of Yahweh turned into an outcast suffocated on an imperial cross. They saw a loser defeated by the powerful, sharing the fate of the persecuted and drowning in the world's tears that so often go unacknowledged. Others saw God in the flesh, outpoured in weakness for the healing the cosmos, with arms wide open in the powerful name of love.

Regardless, Jesus was grasped by the conviction that his pierced hands would continue to unleash an untameable melody of fraternity and cosmic peace, and that somehow, that way, everything would change.

To be sure, the crucified was a stumbling block among the religious and sheer foolishness among the wise. The friend of lepers and prostitutes was a moron by common standards, an "idiot" as Dostoyevsky would later put it.

Yet for those who had ears to hear and eyes to see, Yeshua bar Yosef was the very embodiment of the compassionate presence of Yahweh. According to his earliest friends and followers, Jesus came out alive on the other side of death with a "transphysical" body, to borrow the expression from N. T. Wright.[105] Defying philosophers and scientists, old or new, Jesus' body was reported to have manifested mind-blowing physical properties unknown to what any of us has ever known or experienced, then or now. He went through the muck of death to then come out resurrected on the other side, proving there's more to life than smartphones and popcorn. Against the odds, the central claim of Christianity is that Jesus purged and liberated creation, so that in him and through him all that was and is bruised and corrupted could begin to be renewed and inseminated with an immortality that one day will come into full bloom.

Second-century writer Tertullian of Carthage is often credited to have recognized the eccentricity:

> "The Son of God died": is something immediately credible—because it is absurd. "He was buried, and rose again": that's something certain—because it is impossible.[106]

~

People often live and die eagerly for things they believe to be true; but it's difficult to imagine who would invent the story of the cross and the resurrection to then be willing to die for it knowing it was false, as was the case of Paul, Peter, and the first apostles. But, for all we know, their minds stretched to a seemingly ridiculous point because they *saw and touched* the unforeseen. Because their wild experience outgrew the bounds of their reason, they went on to realize that Yahweh had rebuilt the temple in the most unthinkable way. In a marginal corner of the ancient world's most powerful empire, a small band of once-skeptical *Homo sapiens* met the sort of human that natural evolution and scientific progress have always failed to deliver. In Jesus of Nazareth, they met the victorious *Homo cosmicus* who transcended violence and death, overcoming them once and for all. In Jesus's transphysical body, they said, an inch of the cosmos died and was mysteriously regenerated beyond corruption. They saw and touched the beginnings of the New Creation; they saw and touched the good news for the world, preventing their mouths from being shut and their hands from leaving any stone unturned.

Similar to the days of old when Yahweh had blown the breath of life upon humans, eventually the mystical writer of the Fourth Gospel witnessed to this Jesus breathing the everlasting Spirit into peoples of every race and every nation. Others among his earliest followers recognized themselves as nothing short of living stones of a living temple inhabited by the Spirit of Yahweh. They felt called to be the living hands and feet of the Cosmic Consciousness to whom they felt mystically united. They claimed to be a community of imperfect human beings made and now remade in the living image and likeness of the living God. They experienced the living breath indwelling them, now inspiring them to extend the life of the garden of Eden into every corner of a hurting, rebellious world.[107]

The story of the cosmos began again, affirmed the witnesses of the resurrection. And so they went around turning the world upside down, at times building up a reputation for "acting contrary to the decrees of the emperor, saying that there is another king named Jesus" (Acts 17:6-7).

What might that cosmic renewal imply for an age of global warming when Caesars and their fossil-fueled empires often seem to reign supreme?

5

———Giving to Caesar What is God's?———
Exploring Faithful Political Engagements

> "If we wait for governments, it'll be too little, too late.
> If we act as individuals, it'll be too little.
> But if we act as communities, it might just be enough, just in time."
>
> —Rob Hopkins,
> founder of the Transition Movement

> "Almost always the creative, dedicated minority
> has made the world better."
>
> —Martin Luther King Jr.

FIRE CAN BE LIT for different purposes. It could be set free in an ambush to destroy a village; it could be used as the source of heat in a winter cabin. When let loose, wildfire can be deadly; when used properly, it can warm up and illuminate cold, dark places.

In seeking to respond to the challenges of climate change by looking at the story of the Christian Scriptures, history reminds us that the Christian movement has resembled a fire, leading to both evil and good. When lived out publicly, Christianity has often brought troubling consequences. One needs only to recall again the so-called German Christians who selected portions of the Bible to justify their racial superiority over and against Jews, gypsies, the disabled, and other minorities. The Spanish *conquistadores* also come to mind, having killed thousands with gunpowder in the name of God. There is nothing more contradictory than holding a flag or a weapon in one hand and the Bible in the other—especially when the call of Jesus and Paul was to love one's enemies and "overcome evil with good" (Rom 12:21).

When people are faithful to such a calling, however, Christianity can and has kindled the furnaces of hope and public love. While the call of faith is always deeply personal, the Hebrew-Christian vision has creative reverberations that touch upon every sphere of life. And it is drawing from the inspiration of those who have followed such steps that the second half of this rather long chapter suggests some ways forward to engage publicly in the climate challenge. But not before revisiting the always-sensitive relationship between religion and politics.

"Shining Like Stars in the Firmament"

Against the will of some of his contemporaries, Abraham Lincoln was someone who eventually embraced the apostle Paul's radical declaration that both "slave and free" were united as one "in Christ" (Gal 3:28). Delivering an address in Springfield, Illinois, in 1858, Lincoln asserted that "A house divided against itself cannot stand ... this government cannot endure, permanently half *slave* and half *free*." To be sure, he appealed to logic, to evidence, and to the American Constitution's affirmation that all people were created equal. His speeches and his actions also drew on biblical imagery—imagery he used carefully, not to destroy, but to heal. Instead of abusing his Christian sources to describe the evils of slavery, Lincoln drew from those sources to work out the public repercussions of his faith convictions.[108]

American journalist Dorothy Day also comes to mind, and the worker movement she led, which combined direct aid for the poor with nonviolent direct actions on their behalf. Grounded in the radical vision of Jesus, and in Christian figures such as St. Francis, Leo Tolstoy, and G. K. Chesterton, Day became an advocate of pacifism and of the economic theory of distributism (a theory that allowed everyone to have a share in private property). She devoted her writings and her advocacy toward extending uncompromised solidarity to her fellow workers and the poor around the depression of the 1930s and the decades following. Her Christian beliefs also turned her into an advocate for peace during the Vietnam War, as well as a lobbyist for nuclear disarmament. "Our manifesto is the Sermon on the Mount," she said, resisting the US declaration of war, "which means that we will try to be peacemakers."[109]

It's likewise worth remembering Tommy Douglas, the Canadian Baptist minister who left the pulpit feeling called into the public sphere, where he worked with his political party to eventually introduce North America's first universal health care program. Or recall the more recent efforts of the Evangelical Environmental Network who helped to defend the Endangered Species Act in the US Congress and to organize the Evangelical Climate Initiative based on the moral imperative to take action on climate change.[110]

People and movements like these exemplify the positive ramifications the Christian faith can and should have in public life. They bearwitness to the sort of civic involvement people of faith and congregations can have in laying down a path toward climate justice.

However strangely, sometimes such examples are brushed off as having second-class importance, or as manifestations of a liberal social gospel (the belief that only institutions, and not people, are what need changing). But the brushing-off begs the question: Is there such a thing as a gospel that heals people but leaves the rest of the world unchanged, allowing emperor Caesar to do as he pleases? To better explore some concrete initiatives addressing the climate challenge, we must first step back into the first century to begin to answer that question.

God, Caesar, and a Dangerous Move

When Jesus was asked by the people of his day whether they should pay taxes to the foreign powers of the Roman Empire, he funneled their attention to a foreign coin (a denarius). The Synoptic Gospels then have him replying: "Give back to Caesar what is Caesar's and to God what is God's." (Mark 12:7).

For us who live after the so-called Enlightenment—a time that magnified the split that fifth-century theologian Augustine of Hippo had once made between the "City of God" and the "City of Man"—the saying has come to mean something like this: "Religion and politics are incompatible things. Religion is a matter of personal choice, between you and God; Caesar (the emperor) is about politics and economics." Others would go on to add: "Caesar is in power because God placed him there; therefore, out of obedience to God, obey him always."

Rather tragically, this division between the personal and the political has left many people (Christian and otherwise) believing that today's Caesars can use their coins to do as they please, even if it means wrecking the climate. But can such belief stand on its own feet?

On a few occasions, it's true that some of the writers of the New Testament affirmed the need to honor the ruling authorities (as in 1 Pet 2:13–17). What goes unnoticed, however, is that "honoring" in the sense of esteeming or respecting is one thing, but collusion or blind obedience are entirely another. And that's why the belief that Caesar's power could go unchecked is not entirely fair to Jesus' saying, because collusion is precisely what the religious elites and political leaders of Jesus' day actually *did*: they sided with Rome, even as they disregarded Yahweh's call for peace and justice for all.[111]

Back to the coin and its self-proclaimed owner: Tiberius Caesar, who represented Rome and all the might Rome stood for. He was also an emperor hailed as being the son of a god, the divine Augustus. As such, Rome's imperial propaganda enabled Tiberius to control and tax millions, including Jesus' own people. Hence the rub, for the denarii happened to have Caesar's face imprinted on them. Besides being tools for trade, such coins also served as the ancient world's prime type of mass media. They were used as megaphones of sorts to amplify the emperor's wealth and power to put entire nations under his feet.

So put yourself in the sandals of first-century Jews whose people continued to be crucified by the Romans, who learned from Moses that no images should be engraved anywhere, and who believed no foreign master should ever be served except Yahweh: being forced to use Caesar's coins and pay him tribute was close to a slap in the face.[112]

Various movements of rebellion arose in first-century Palestine in response to the compromise. Protests in Jerusalem against the Roman and Jewish elites bubbled time after time, fueled by the memories of the Exodus and by more recent revolts, such as the one led in 164 BCE by Jewish priest Judas Maccabeus against the invading Seleucid Empire. The movements, in turn, often called for armed revolution against the authorities: "Give Rome what Rome deserves! Rome has to be paid in full, eye for eye!"

For common folk in Israel, the compromise was an offense. In fact, it was blasphemy. As good monotheists, they knew that by siding with

Caesar's empire—as Herod Antipas and the temple's chief priests were doing in one way or another—the entire nation was failing to worship Yahweh as they should. The political and religious elite of Jerusalem was, in fact, colluding with Caesar, indirectly giving him the loyalty that only God deserved. Pushing back, zealous rebels advocated for violence—even if, however ironically, that meant using their enemy's wicked means to defeat it.

This turbulent backdrop leads historian N. T. Wright to dismiss Jesus' famous God/Caesar saying as a blank check allowing Caesar to have his way as he pleases and people to obey as he commands. Instead, Wright hears a three-sided *critique* in Jesus' words: a critique of the profane claims *of* Rome, a critique of the Zealots' call to violent revolution to *defeat* Rome, and a critique of the collusion of the elders and religious leaders of Israel *with* Rome.

Jesus cheered for none. Collusion with Rome was wrong, but so was violence. The call of the Nazarene, Wright suggests, was for a deeper revolution: one of peacemaking and of imitating Yahweh's self-giving love. Aimed for those with ears to hear, Jesus' words were subtle but highly subversive: "Give that dirty coin back to Caesar, and the dedication and loyalty you've been giving Caesar, give it to God."[113]

Calling for a very different sort of revolution, Jesus challenged his contemporaries to be the light of the world, not a reflection of Rome's empire. But in doing so, he did not put out a crippled call for eternal souls or isolated individuals searching for a self-absorbed religious experience. Jesus' three-sided critique was a public statement; it was a call to embody Yahweh's longstanding dream of building a society of love.

The Head of Governments and Corporations: An Impolite Interruption

Back to the 21st century.

Traveling in time from Jesus' day into ours, some of our friends from chapter 1 now make an impolite reappearance. Rubbing his eyes at the sight of the scene, John the seer stands bewildered at our condition, quick to note how dethroning Yahweh in the name of profit has in fact allowed new powers to take over society's throne. He hears of corporations such as Shell, BP, Amazon, or Microsoft being praised as monster companies, even as he witnesses a few individuals being hailed as the new cosmic priests capable of unlocking the mysteries of corporate success.

Unable to cry, John smiles in irony describing what he sees. For all he knows, despite Caesar's claim to the throne, the world only has one chief ruler. But there they are: wrapped in neon lights, national governments and corporate bullies are now claiming center stage. The first-century Book of the Apocalypse suddenly becomes relevant for the twenty-first.

"But Jesus said his kingdom was not of this world!" someone objects.

"Who cares about the planet... we're off to heaven soon!" adds another.

"It's all going to be burned up, anyway!" yet another one remarks.

"Of course it's not of this world," John cries out loud. "In the kingdoms of this world, rulers are into money and power at all costs. In the kingdoms of this world, Caesars want to be served by many, revered by most."

"Indeed. It's not *of* this world, but it's *for* this world," adds one of the Floating Clowns, making a serene arrival with a smirk on her face. "The ecosystem of heaven is meant to heal the one on Earth here below."

John leaves little room for linear argument. With the flare of a poet, the seer enlivens the imagination with candid metaphors. Transcending the bounds of reason, he goes on by reminding his hearers how the Lamb did not come to be served, but to serve and give his life in the service of many. The contrast becomes stark once he pinpoints glitzy corporate beasts running around unleashed, wanting to take over the cosmic gala. Their power lures all peoples of all nations.

But the seer's vision cuts through the fake atmospherics and moves on to disclose what's taking place backstage. "The beasts are political and religious powers that are ready to deceive many," John says. "They're puppets at the mercy of the dragon, the ancient serpent of old. The beasts and the dragon are the unholy trinity, the ultimate parody of God."[114]

In contrast, the exiled visionary refurbishes the imagination. Not a beast, not a lion. John sees a humble Lamb on a throne: the King of kings clothed in light with a golden sash around the chest. He also sees a great multitude that no one could number; people from every nation, from all tribes and all languages. They stand before the throne and before the Lamb, all clothed in white robes.

In white robes. The people had washed their robes and made them white in the blood of the Lamb. Made white, in the blood. White as a symbol of their pure commitment to follow the Lamb wherever the Lamb went. White because they did not shy away from sharing the Lamb's unglamorous destiny. White because John knew that the King of kings has the blood of a lion, but the character of a lamb: fierce, but meek and humble in heart; slow to anger, abounding in unwavering love . . . a power so enormous and so noble that it has always been a threat to the world's Caesars and to all who follow their steps.

∼

What would it look like to "follow the Lamb wherever he goes" (Rev 14:4), especially now that King Coal and Prince Oil have become self-obsessed powers of beastly proportions standing behind the reality of climate change? What would it mean to honor Yahweh as people, as communities, and as cities in a world that today's Caesars continue to claim as their own? The remaining of this chapter is geared to those readers with a more analytical or strategic personality; less technically-oriented readers are welcome to jump to the section "Citizens of Heaven Rooted on Earth" a few pages below.

Practical Climate Responses at All Levels

Back to a few telling stats. Visualize fuel for a moment: at any given time, every car in the United States carries approximately seven gallons of gasoline. For every mile traveled, an average car that runs twenty-five miles per gallon spews out over forty-five gallons of CO_2. This pollution goes into the atmospheric blanket, which has continued to warm since the Industrial Revolution. Accounting for all vehicles, that's 400 million gallons of gasoline entering and leaving the car flow every day—literally, rivers of gas flowing across roads and highways.[115] Canada's energy panorama is similar. Even if on a national level Canada's contribution to global GHG emissions is less than 2 percent, those of us who live here have one of the highest carbon footprints worldwide (twice the yearly average emissions of Europeans). On average, we're burning around twenty-three liters (six gallons) of fuel per person per day.[116]

Can we shrink these footprints and make this river of gasoline run dry by the sheer sum of individual efforts?

"Fair-Trade Maple Syrup, Madame?": Responding as Individuals

Someone once said that all in favor of conserving gasoline should raise their right foot. Valuing all sorts of trees wherever they sprout, it's important to value meaningful action on every front. Personal-level responses to the climate challenge are necessary and ultimately inescapable. Even if they seem insignificant, our buying choices and personal lifestyle changes do have a role to play. Not in vain are we reminded that by our fruit we shall be known and that people of faith should be motivated by love and not always by circumstance. Personal changes and standing up for what is right are important regardless of the outcome.

It's also the case that focusing only (or primarily) on individual efforts is not enough. Even if many small changes in our behavior were to contribute toward a greater difference, our societies are designed in a way that makes these changes insufficient. For one, only wealthier sectors of the population can afford so-called green products. As the saying goes, "The rich get organic food; the poor get diabetes." For another, individual changes can only add to so much when there are much larger powers at play. Consider freshwater use and waste. Second highest on a per capita basis relative to 29 wealthy nations, household consumption in Canada accounts for 20 percent of water use; industry is responsible for 68 percent.[117] In British Columbia (arguably Canada's greenest province), 37 percent of waste disposal comes from households; 63 percent is chalked up to non-household sources.[118]

The burning of fossil fuels is similar. Even if all of us have a share in warming up the planet, between 1988 and 2015, just 25 fossil fuel extractors were linked to 51 percent of global industrial emissions, while 100 of them were responsible for 71 percent of all GHGs.[119] Likewise, while people like me living in Canada do have one of the largest carbon footprints worldwide, 49 percent of those emissions result from the transportation systems (23 percent) and the oil and gas (26 percent) sectors alone.[120] These are two realms over which we, as isolated individuals, have little personal control. The greater burden of responsibility falls on polluting powers that are beyond our personal level of influence. So, important as they are, it's simply not enough to plant a tree and ride a bike to save the world. Individualizing the problem that way doesn't even *tickle* the bestial power (Rev 13) of fossil fuel corporations.[121]

More to the point, believing that the burden of responsibility falls primarily on individuals alone (or on buying) actually reinforces our societies' two prevailing myths: neoliberalism and consumerism. By thinking we should merely "vote with our dollar" as a way to effect change (or as *the* main way to effect change), we actually keep oiling the wheels of a supposedly free market system. But that leaves the sinful causes of the problem untouched. Buying "green" still allows for imposing one culture upon another and for endless economic growth at the expense of whatever. Such system is fueled by choice, greed, convenience, and self-interest, which are values at odds with the call to brother and sisterhood, justice, and the self-giving love that is central to the Christian way.

Fueled by the blind faith in market fundamentalisms regretted by American economists such as Joseph Stiglitz, the neoliberal mythology has in fact done a good job of portraying our societies and human communities as nothing more than a rambling lump of selfish and self-contained individuals. Consciously or not, this propaganda champions a new spin-off of the "entertain, divide, and conquer" mentality that has prevailed at least since the time of the Romans.

Calling for a more audacious embrace of sustainability, reasons like these have led renowned international business consultant Wayne Visser to conclude that if

> we want to achieve scalability of sustainable and responsible products and services, we cannot leave it to the passive choices of customers.... Companies [and governments] have to show their willingness to set bold, audacious targets that will reverse the negative social and environmental trends.[122]

What might that look like?

Congregations as Hubs for Action: Community Responses

There are an estimated 350,000 church communities in the United States, and 21,000 in Canada, which combined gather around seventy million congregants. Not that numbers ultimately matter, but imagine if a fraction of these communities committed themselves to taking thoughtful and compassionate action to address climate change. That could create a groundswell and lead to a decentralized revolution sprouting all over. Following the nineteenth-century abolitionists or the twentieth-century early civil rights advocates, what would happen if the pulpits stretched their reach to include climate change alongside poverty, child abuse, or racial discrimination? Is the time not ripe for priests, ministers, pastors, and spiritual leaders to take courage and exemplify what it looks like to respond collectively to the situation? Caring for our neighbors, after all, is at the core of what Christianity has always been about.

Churches also have the great potential of becoming hubs where people can be inspired to collaborate toward bigger goals that escape the short reach of individuals. In our virtual age of isolation, faith communities can bring people together and help reweave the torn fabric of our virtual societies. Plus, doing good is actually contagious. One is more likely to purchase fair-trade produce or use public transit if one's friends do so too. Churches could take collective pledges as a way to be mutually-inspired and hold each other accountable. Means allowing, these pledges could include guidelines for personal life (what foods to eat and where to source them, modes of transportation to use, energy-saving practices at home, switching to renewable energies, etc.), but pledges could also be embraced by churches collectively (as much as possible, by sourcing their food directly from virtuous farms, promoting bikes and transit to attend weekly services, removing institutional investments from fossil fuel industries, and so on).

By committing collectively, congregations can both challenge and inspire their members to embrace smaller personal changes that, in the end, add up. Collective actions also make for a good institutional story, which can send ripples through the media and through one-on-one conversations—all necessary elements to inspire a climate that leads toward wider change.

Space preventing examples of churches taking action on climate change, here readers are left to discover practical guidelines and stories of inspiration from A Rocha's "Eco Church," the Office of Social Justice's "Climate Witness Project," or Faith & the Common Good's "Faithful Footprints."[123]

From 9 to 5, Monday to Friday: The Workplace

As important as it is to "green" a congregation, we spend most of our productive time in the workplace. What would it look like to move toward a vision of climate justice in the marketplace and at work? In what ways can each of us sow and cultivate our contributions, so they grow into a transformative atmosphere in the companies, institutions, and organizations that we're part of?

While it's difficult for someone alone to effect change in their workplace, a committed group of like-minded eco-champions has a better chance of advancing meaningful initiatives. "Green teams" are a promising way forward when working out possibilities such as divestment, composting, responsible procurement, closed-loop servicing, switching to renewable energies, and the like.[124] These teams can brainstorm different proposals and working plans that can be tested and eventually presented to directors, outlining a clear set of organizational benefits.

Here the "Transformational Company Guide" offers various case stories and bold, innovative paths of action toward circular, inclusive, low-carbon business models.[125] Equally important, green teams are challenging and motivating their companies to partake and advocate for low-carbon coalitions already underway. One of the most promising business coalitions in the US is "Business for Innovative Climate and Energy Policy"; in Canada, companies could join and support the "Pan-Canadian Framework on Climate Change and Clean Growth."

Calling Out the Elephant in the Room: Necessary, but not Enough

For all their worth, company- and church-based responses also have some shortcomings. For one, climate change and ecological disruption are both driven by economic growth—even by so-called clean economic growth. For another, some church services have become restaurants of "Chicken Soup for the Soul." Congregations often look more like social clubs or therapy centers but remain largely disengaged from the land, air, and water they share with countless other humans and non-human creatures. Equally concerning, many of us Christians today are no longer accustomed to thinking (let alone acting) as a collective—as a body. Asking congregants to sacrifice their living comfort will not ring a bell with some pastors, because their salary usually depends on the tithing and offerings of their flock. If the pulpit is too demanding, or too outlandish, many churches run the risk of losing at least part of their congregants. People could leave searching for an easier way, for sermons more aligned with their choices and music more attuned to their ears. (Some have described this consumerist takeover of the message of Jesus as the "McDonaldization" of the church: *Double-sized blessings, Mr.?*)

Narrowing efforts to the community level can also overlook the bigger picture. This became clear to me when I helped to assess the carbon footprint of my former faith community in Vancouver: Grandview Calvary Church. Considering greenhouse gas emissions from waste disposal, heating, electricity, and staff transit, a friend and I tracked the institution's carbon footprint (excluding what the GHG Protocol calls "scope-3" indirect emissions, which account for emissions outside of an organization's direct control). We found out that the yearly consumption was equivalent to burning twenty-four barrels of oil every year. But we made a comparison. When squared against the intended expansion of the Kinder Morgan pipeline (a pipeline very close to Vancouver, which would increase crude exports from 300,000 to 890,000 barrels/day), the twenty-four barrels/year paled in comparison. So we realized it would be around 8,970,000 times more impactful to join our energies with other municipal and citizen groups to stop large-scale energy projects, instead of simply focusing on reducing one building's carbon footprint (important as that is to do).

Disproportions like these explain why citizen organizers urge us to be more audacious. Annie Leonard from "The Story of Stuff Project" admits that "[personal] actions are a fine place to start. But they are a terrible place to stop." Building on historical case studies such as the civil rights movement, she calls us, instead, to work for broader societal change by embracing at least three things. First, a big idea of how things could be better. Second, a commitment to go beyond individual actions. And third, beyond tweets and posts and stories of "individual eco-perfectionists," we need a collective engagement for political and structural change.[126]

Climate Change, or System Change?

That is an enormous question with even larger implications, but consider that the World Economic Forum estimates that, in the transition toward a future of low-fossil fuels, every $1.00 not invested today would cost us at least $1.75 (and up to $4.00) a few decades down the road. If we feast on fossils fuels now, our hangover will impede climate justice later.[127] To avoid the more drastic effects of climate change, our neighborhoods, companies, and cities need to shift away from our current system based on growth and endless extraction, toward one designed for well-being and permanence. The clock is surely ticking for low-energy systems to be birthed out of the womb of the old now.

Leaving behind our fossil-addicted system is no small challenge. The path toward climate justice calls us to phase out long-term, infrastructure-heavy energy projects rapidly. Pipelines, coal plants, and LNG facilities need to go sooner than later. In fact, in high-income nations such as the US and Canada, GHG emissions should essentially decrease almost to zero come 2050—around 30 percent each decade in the next thirty years. And as the now-famous documentary *This Changes Everything: Capitalism vs. The Climate* has demonstrated convincingly, that's not going to happen with free business as usual or magical techno-fixes. Pushing government regulations out of the picture is what got us into the problem in the first place.[128] Instead, governments should be very sober in using what ecologists call our remaining "fossil fuel budget" in a coordinated and steadfast transition toward policies and urban redesigns that stop taking cheap energy for granted.

Courageous Leadership from Within (and from Without)

While diplomacy and negotiations have a place, more than twenty-five years of slow-moving international climate treaties have shown that national governments are far from being the key leaders in the energy transition. In fact, they often stand in the way. Simply consider the federal powers of Caesar's whim, which are close to pulling out the US from the 2015 Paris Accord, a non-binding international pledge signed by over 170 nations committed to significantly cutting down their GHG emissions.

Despite the baggage and compromise of big governments, we still need groups of committed citizens to be engaged in everyday politics. Voting for enlightened governments can play a small part in the way forward, as long as leaders are bold and act with wisdom, compassion, and integrity. Whether from presidents, politicians, mayors, or CEOs, courageous leadership is needed through and through.[129]

On this hope, the faith-based NGO Citizens for Public Justice proposes three key, urgent political pathways that decision-makers should pursue and citizen initiatives can lobby and advocate for. These pathways need to consider the economically less-advantaged through and through, such as working-class suburban commuters that have to travel long distances to work, or workers currently employed in the fossil fuel industry. With that in mind, we need long-term energy plans that:

1. Put a substantial price on fossil fuels (called a "carbon tax") to de-incentivize their consumption (and to account for the real cost of the

damage of climate change), as well as strictly regulate carbon-intensive, extractive sectors such as forestry, coal, oil, and gas.

2. Simultaneously develop a low-carbon economy by gradually but steadily eliminating subsidies to fossil fuel sectors, while subsidizing low-carbon initiatives and urban redesign.

3. Provide justice for the poor and disenfranchised through social supports, transition jobs in the circular economy for people currently employed in the fossil fuel sector, and fund the UN Green Climate Fund (a pool of money destined to support low-income nations prepare for a low-carbon future).[130]

For all the worth of these three proposals (towards which we should work tirelessly), Boris Waterlove reminds us that, even better than cleaning the water downstream, is avoiding polluting it upstream. Put bluntly: the central problem of climate change is not GHG emissions, or carbon footprints, or lacking solar panels. The problem is the toxic blend of our ever-hungry economic system and our endless *extraction* of fossil fuels. Once fossil fuels are pumped out of the ground, the temptation to burn them is simply too great.[131]

So the three recommendations above must be implemented but also outgrown by this slogan: *"Fossil fuels must be steadily phased out starting NOW, and that means leaving them in the ground where they belong."* And here's the rub: Governments and corporations, left to themselves, *will not make that happen.* Only with steadfast, long-term, collective citizen pressure is that likely to be the case. And to build such positive pressure, the climate movement needs to join efforts with the people most affected by climate change—workers and labor groups, indigenous peoples, and all the other visible minorities who have reaped the negative side-effects of the triumph of the West.

Similar to the times of the abolitionist campaigns, churches can come side-by-side with these groups and seek to honor the steps of Jesus who embraced the destiny of the no-bodies. People of faith and of good will can give themselves as volunteers, cheer for these groups as advocates, support their work financially, or even better, leave behind their Wall-Street-ified jobs to work full-heartedly toward a renewed economy.[132]

A Climate of Desire

Of the People, by the People, for the People (and the Land)

It's worth recalling that for the past 4,000 years or so, new empires have always risen after others have fallen. History and the story of Babel remind us that we should be suspicious of real change coming by leveraging the existing systems of power. In fact, it probably won't. While we should hope and work tirelessly for structural reform, renewal is much more likely to come when alternative, committed minorities operate in the system and challenge the system, but without being part of it (minorities that are "in the world" but not "of it").

In his classic *When Corporations Rule the World*, former Harvard professor David Korten documented how today's elected national governments can be at the mercy of short-term, corporate interests. That creates a problem every 4 or 5 years when short-sighted political parties get hungry for votes, often to serve the interests of a Wall Street's fossil-fueled corporate machinery that endlessly craves quick profits. The short-term tends to win because, save for a few exceptions, corporations themselves are impersonal entities *designed* with a "pathological" hunger for profit and power, to borrow the term from UBC professor of corporate law, Joel Bakan.[133] Or to recall Jeremiah: When they fail to serve the common good, governments and corporations behave like self-obsessed idols that have eyes that do not see, and ears that do not hear (Jer 5:1; Ps 115). By bowing down to money and power, these modern giants can easily "prostitute themselves with many lovers" and go on "defiling the land with their vile harlotry" (Jer 2:7; 3:1).

In response, we *can* stand collectively with inclusive, citizen-led initiatives. In contrast to Babel's ways of domination and centralized power, change has been emerging from the underside, from civil movements and organizations embracing a kind of "decentralized robust action"—initiatives run by the people, for the people (and the land).[134] Recall here the recent grass-roots citizen campaigns in Hamburg, Germany who won the vote to take back their electrical grid from foreign utility company Vattenfall and start decentralized energy cooperatives instead. Inspiration is also found in Bolivia's water struggles in Cochabamba and La Paz, which spurred local "remunicipalization" efforts that took water away from private interests to democratize it to the public.[135] And on a more sobering scale, we are always humbled of course by figures like Gandhi leading India into independence from Britain's yoke, or Martin Luther King Jr. spearheading the civil rights movement. Inspired by the example of Jesus, both Gandhi and Dr. King created movements that operated outside of the formal structures of power, and working at the grass-roots level enabled them to achieve far more than any traditional leader could ever aspire to attain.[136]

∼

What are some promising citizen-led initiatives to support? In the United States, climate-related initiatives of this sort have been put forward by the Climate Action Network and the Citizens Climate Lobby. The Dogwood Initiative, Courage, Indigenous Climate Action, and the Canadian Youth Climate Coalition are worth supporting in Canada. Likewise, "The Leap" is growing as an international movement for just transition away from fossil fuels, integrating a move to renewable energy with livable incomes, affordable healthcare, respect for indigenous rights, and more. Under the slogan of "People Powered Change," these initiatives have been taking on a slew of issues from electoral reform to pipeline resistance to renewables investment.

To be sure, initiatives like these are imperfect and prone to corruption as any other human initiative, but they do tend to have a long-term view, serving the real interest of people and places. In contrast to the corporate-driven or short-term whims of elected governments, people-focused initiatives have a better chance of working toward intelligent, enduring transition plans. In fact, because most of us can hardly achieve much nationally (let alone globally), more and more ecological aficionados agree that *local collective actions* and *unco-optable social movements* are perhaps the most promising way to change "the system."[137] And they remind us of an old truth that our technological society of mastery and domination has made us forget: We are humans, not gods, but that's precisely why we *can* act within our municipalities and our watersheds. We are to be primarily concerned about cultivating forms of life and taking a stand *within the location and place where we live*. System change will more likely come about when a dispersed groundswell of countless groups, churches, and faith-communities take an active stand and look after the people and places each of us inhabits—a vision not far from that of the Book of Genesis, which summons us all to spread and exercise care and humble dominion, instead of gathering up around our mighty towers of Babel to have centralized control and domination.

Besides cheering and supporting these initiatives in every possible way, how else might we begin to outflank Caesar and be loyal to God by embracing Jesus' all-encompassing renewal of cross-bearing, self-giving love?

Citizens of Heaven Rooted on Earth

As citizens of heaven summoned to work for the flourishing of our earthly home (Phil 3), practitioners of the Christian faith have a great opportunity to partner with all sorts of people of good will in the healing of our cities. Regardless of the whims of empire and Caesar's edicts, we can become agents of renewal in the places we dwell, helping our cities and institutions break free from their urban addiction to cheap carbon cocktails.

Having sojourned into a renewed world alongside many people who are working tirelessly toward a future with low-fossil fuels, allow me to share what the last years have taught me in organizing practical responses to our current crisis. Under the motto "Don't reinvent warm water, but jump into the thermal hot springs that are already bubbling," consider at least three major collective transitions faith communities should support—be it as fans, volunteers, advocates, workers, or funders.[138]

Divesting from (Pipe)lines and Investing in Cycles

The city of San Francisco, the World Council of Churches, the Canadian Medical Association, Stanford University, and the First Presbyterian Church of Florida are a few among the 890+ universities, religious organizations, municipalities, and pension funds around the world that have taken up the global call to "divest," to pull their investment funds away from the coal, oil, and gas industries.

This campaign has come to the rescue as a boycott of sorts, chipping away the disguise of fossil fuel companies that continue to clothe themselves as angels of light. Instead of reaping from an industry that is wrecking the planet, divestment allows church communities to put their money where their mouth is. According to former Anglican bishop Desmond Tutu, the math is simple: "It makes no sense to invest in companies that undermine our future. To serve as custodians of creation is not an empty title; it requires that we act, and with all the urgency this dire situation demands."[139]

Divestment will not fully address the climate challenge. But it will help, just as it has in attaining other social victories. Backed by churches, divestment strategies were deployed to expose the segregation of black people through apartheid in South Africa. They were also used to create broader public awareness about the harmful effects of the tobacco industry. Likewise, today's divestment movement is sending a clear message through the public megaphone by continuing to create ripples of awareness, while exposing the beastly behavior of some governments and fossil fuel corporations.

In fact, according to the global Fossil Free campaign, churches and faith-based institutions worldwide are topping the list with 30 percent of the divestment pie, which keeps growing beyond its current level of $6.15 trillion USD.[140] Besides the World Council of Churches, other churches include the Episcopal Church in the USA, the Church of Sweden, the Church of England, and the Lutheran World Federation. But dozens and dozens of small parishes and congregations are also contributing their grain of sand.

To be sure, divestment could still perpetuate the bottomless bellies of financial systems like Wall Street's. Hence the energy transition calls for re-investing in decentralized, citizen-owned and cooperatively operated energy co-ops that prioritize energy retrofits in buildings as well as regional renewable energy grids. Consider a few questions to ask when seeking to support citizen movements or to reinvest in renewable energy operations:

1. Is the initiative prioritizing urban reconfigurations that cut down and eliminate the need for energy consumption *in the first place*?

2. Is priority given to mass public transit run on renewables and to building high energy efficiency into the existing built environment?

3. Is the alternative energy operation democratically owned and operated (for example, citizen co-ops, social enterprises, community interest corporations)?

4. After reconfiguring urban spaces and cutting back energy consumption, is priority being given, first, to onshore wind turbines, and only secondly, to solar farms, solar rooftop, and geothermal?

Eating in Circles: from Agriculture to Permaculture

As Wendell Berry made it clear in "The Pleasures of Eating," most of us eat in great ignorance today. We bypass that the industrialized food system is built on a linear, mechanical logic that abuses animals and that is blind to the rhythms and cycles of the land. We sidestep the deforestation and the copious amount of fossil fuels we're using to grow, move, and store our food. We overlook the paving of prime agricultural land with buildings and parking lots, even as we degrade it with synthetic chemicals and often with careless practices that pollute watersheds and erode the top soils that nourish us.

Thankfully, the last few decades have witnessed a surge of lovers and practitioners of every kind—people who care intimately for their farms and their places. New modes of sovereign, ecologically wise food production such as agroecology and permaculture have been taking root, even as more local supply chains and new bonds of direct trade have been developing between food growers and food eaters. Farmer's markets, food hubs, and direct fair-trade buying clubs are very promising alternatives.[141]

Still, these ways of feeding ourselves usually fall only within the reach of wealthier sectors. What we now need is collective citizen support for civil initiatives that aim to debunk Babylonian supermarkets, even as they seek to create ecologically regenerative food systems, while minimizing food waste. The goal is that a primarily plant-based diet becomes accessible to all, using the least possible amount of fossil fuels. Here, then, it's worth supporting the North American chapters of the *Vía Campesina*, namely the National Farmers Union or Food Secure Canada (in Canada), and the Rural Coalition and the National Family Farm Coalition in the USA.

The act of eating is one where sex, Christianity, and climate change come closely together. If we see it as a gift (as it has been seen for millennia), the eyes of gratitude can once again reveal the sacredness of food. We live through it. Besides the water we drink and the air we breathe, food is our most intimate bond with the rest of the created world. Perhaps as intimate a sex. Food goes "in" us; it becomes part of us. And that's why when we give proper consideration to where our food came from (who grew it, under what conditions) our tables have the potential of becoming a place of bonding and communion between us humans, our living source, and the creatures and the land that sustain us. An intimate way of interweaving creatures and creation, eating well is in fact a holy act of sacred intercourse.

Living in Circles: from Shopping Malls to Garden Cities

Last but not least: the religion of consumerism and entertainment must be left behind. These myths must go. As national governments have grown fat, and corporations even fatter, municipalities and citizen groups have gotten slim. In fact, we hardly use the language of citizens anymore. We speak of ourselves as employees, we refer to countries as markets of X, Y, or Z million consumers. As corporate banks continue to lure us and dictate the norms and whims of our cultures, we have come to see ourselves as little more than serfs at their mercy. In turn, shopping malls have become our new cathedrals and cinemas our new amphitheaters. We're stuffed with entertainment, all dressed up for Lady Gaga, but bored, stressed, and with nowhere to go once the virtual fireworks are over.

But what if entertainment doesn't need to be abandoned, but reoriented? Instead of consuming our way into social unrest and ecological exile, this is our opportunity to get creative by living with less but enjoying it more. Some today call this voluntary simplicity; our grandparents knew it as homesteading and neighborliness. We can turn our kitchens and living rooms into places of creativity and teamwork, places to grow deeper friendships around food, games, and music. Whether making bread, beer, or kombucha, the journey from consumerism toward sustainability calls us to rediscover our hands by actively designing, remaking, and fixing stuff instead of passively craving the latest gadgets by shopping until we drop.

This may sound idealistic, or regressive, or uncorporate. Well, it is. And it may also sound like sliding back into the me-centered world of individual change. But that's only partially true, because a key ingredient of system change is people and small committed communities exemplifying what it means to transition toward a low-energy, circular economy. Simplicity creates monumental ripples and a creek always flows into a stream, a stream into a river, and a river into the endless ocean.

But this sort of change is not entirely personal. Led by compassionate and enlightened leaders, churches and citizen groups can support municipal initiatives to strengthen local social enterprises and workers co-ops geared toward the well-being of our neighborhoods and cities. Bike rental and repair shops; furniture refurbishing; home and building deconstruction hubs; edible tree initiatives; centers for tool sharing—these are only a few examples. Similar to relocalizing our food systems, we also need to stop exporting jobs and pollution and instead bring industries for daily necessities closer to where we live. Groups and churches can support long-term projects that redesign cities and economies to stop taking cheap fossil-fueled transportation for granted so that at the end of the day most roads remain local.

Here the Transition Network is a valuable worldwide movement to tap into and support (www.transitionnetwork.org). Readers are encouraged to gather with like-minded people and explore which initiatives are taking place in their particular location.

∼

In his classic *Small is Beautiful: Economics as if People Mattered*, the above-mentioned German economist E. F. Schumacher once recognized the need for high-income nations, not to maximize consumption, but to maximize *satisfaction* with a minimum of consumption. Quite simply, our planet cannot hold that many of us consuming that many things. Further, it has been widely documented that, beyond a certain threshold, the act of buying does not actually make us better off. "There was a man so poor, so poor, so poor," the saying goes, "that the only thing he had, was money." Somewhat of a caricature, but true for the most part.

It's rather the sharing of positive experiences that gives life. Many of us don't need more money; we need more time, more peace, better and deeper friendships. We need to rediscover the art of being engaged citizens and welcoming neighbors. And, far beyond, we also need a new attitude. Giving back to God the coin and the worship that Caesar has claimed as his own requires an altogether different posture. For the climate to truly change, we're called to stop behaving like masters of consumption, and become, instead, practitioners of earthkeeping.

Is there hope for such a journey?

6

When the Climate Changed
Hope for Today

"Only when Christ comes again will the little white children of Alabama walk hand in hand with little black children."
—Billy Graham

"It is wonderful how much time good people spend fighting the devil. If they would only expend the same amount of energy loving their fellow men, the devil would die in his own tracks of ennui."
—Helen Keller, American Author

THE MAIN SEED FOR this book was sown in 2013. I very well remember the sunny Monday morning sitting at my work desk at the Asociación Empresarial para el Desarrollo, a local hub for business responsibility in Costa Rica. About two months into my new job, I had began looking into some of the local conditions related to society and the economy. As my eyes wandered, staring through the window, eventually a ray of sunshine found its way through the blinds, bringing my attention to two regional reports sitting not far from me. Curious, I took a look.

One of them, "Vision 2050," was issued by the World Business Council for Sustainable Development, a coalition of 200 of the greatest corporations in the world. To my surprise, it argued that we can't go on with business as usual. And since no one I know ever admits his or her faults just for nothing, I quickly realized that there had to be some truth to the matter because it was corporations themselves that were confessing that the business world was failing. The report went on affirming that we will soon require three planets to sustain our levels of consumption. It also emphasized that, because of our burning of fossil fuels, over the next decades the average global temperatures are expected to rise between 4 to 6 degrees (although more conservative estimates say the elevation could reach anywhere between 1.5 and 4.5 degrees). Soon after, I came to realize that the risks of climate change included severe water stress, floods, crop failures, sea-level rise, mass migrations, and political unrest.[142]

Those were all distant threats, I thought, but as I read on, I discovered that they were not so far away. Costa Rica borders to the north with Nicaragua, and between Honduras and Nicaragua, the other report stated, several million people were likely to be displaced as climate refugees—and that meant Costa Rica, that meant me.

Displaced peasants and poor workers have been migrating to Costa Rica for decades. But finding increasingly less work in the countryside, they have begun to settle in the outskirts of San José, the nation's capital city. In turn, this influx has created what is called "rings of misery" (urban peripheries that breed and spread poverty, drug dealing, and crime), translating into deplorable living conditions and social unrest for all.

That was all sad and disquieting enough. Still, I became particularly uneasy because the report went on to say that, even if responsible for less than 1 to 2 percent of the world's annual GHG emissions, the Central American region is considered the most vulnerable tropical "hot spot" on the planet. Even if climate change is affecting us all (and will affect us all, rich and poor, north and south), scientists have targeted Honduras and Nicaragua as the 3rd and 5th most vulnerable countries, globally. They are among the top of the list of the billion climate refugees estimated by the end of this century.[143]

Ever since I've asked myself: "Where is the God that Jesus revealed in the midst of this?" "Is God dead," as Nietzsche declared? "Is God detached and absent—the opposite of a control-freak god—giving us humans so much freedom to the point of allowing injustice to go on unchecked?"

I was also confronted by a disruption in my chest of whether I should care about climate change selfishly, so to speak, only because it's going to affect me, or whether I should care because the world isn't mine, because

I have a responsibility toward God, toward others, and toward the living world itself. My own selfishness inclined me toward the former; but beyond the virtual cloud of fear and despair that tempted me to not look anywhere beyond the round tip of my nose, my deepest convictions have continued to suggest differently.

Angels & Daemons

The New Testament authors spoke about the coming of the kingdom of God (or the kingdom of heaven) both as an event at the end of history, but also as a living reality that Jesus already inaugurated. It is not, however, a kingdom up in the skies, unrelated to what goes on here below. Nor is it a kingdom of a heavenly dictator sending down lightning bolts to punish humans. For Jesus and the apostles, the coming of the kingdom meant the come-down of heaven's clean-up operation. The kingdom is the restoration of relationships, the establishment of harmony and flourishing in every corner of the world—in ourselves, with our neighbors, with our living source, and with the rest of the living world. The kingdom is the ultimate ecosystem where everyone and everything thrives. The ancient Hebrews spoke of it as *shalom*; John the seer saw it as a kind of sacred intercourse between God and all things on heaven and earth; one of today's faith leaders calls it the "civilization of love."[144]

A Blinding Light

This kingdom dream is far from being realized. As it should be fairly evident to all of us, every person and every civilization we have come to know of so far has been wounded by evil. And despite our best intentions and our best efforts to curb it, our societies and the living world have always called for a healing that remains beyond our reach. In fact, our best intentions for solving our problems often create even larger ones: solar panels require tons of heavy metals and are loaded with poisonous chemicals; synthetic fertilizers help produce cheap food, but they affect the health of the soils, pollute watersheds, and require plenty of fossil fuels to be manufactured. Technology often enables us to take one step forward, but only by taking two (or three) steps backward.

This irony has a backdrop. The most obvious one is that our zeal for tech fixes continues to reflect our subtle but pervasive domination of nature. Technology can easily set us further and further apart from anything else that is alive. But perhaps most significantly, our technological society continues to be a significant consequence of setting ourselves as humans at the center of all things. Some Italian humanists of the Renaissance made us believe that our role is to have a vertical relationship with the living world without any sense of horizontal reciprocity toward it. One of the fathers of modern science in the sixteenth century, a British man by the name of Francis Bacon, went further by affirming that the goal of science was to achieve progress by implementing what he saw as a God-given call to dominate nature through research and technology. He thought that science should put nature on a rack and torture her secrets out of her. Others like René Descartes made us believe that the world and people behave like machines, that animals have no feelings, that the mind and the body are separate things.

Tragically, both Mr. Bacon and many of the humanists of the so-called Enlightenment grounded their ideas in the call of the Book of Genesis to have "dominion," a call they distorted and misinterpreted as if it was a blank check for humans to exercise domina*tion*. Equally concerning, Mr. Bacon believed that the living world was created solely for the sake of humanity, and, as such, he aspired "to stretch the deplorably narrow limits of man's dominion over the universe to their promised bounds." He went as far as urging his contemporaries to "bind nature to our service and make her our slave."[145]

Ideas like these turned folk like him into the chief architects of our modern towers of Babel. Likely with noble intentions, thinkers like Descartes and Bacon helped create the cultural push and the intellectual map that eventually gave the West full license to set itself over and above the rest of the living world at full speed.

Equally tragic, this quest for domination eventually came at the expense of crushing and neglecting nature's spiritual dimension—often to the point of leading us to believe (religiously) that matter is all there is. And often to the point of making us oblivious to the invisible forces and realities that shape our cultures, whether angelic or demonic. And such neglect calls for a quick visit to times past, going from the seventeenth century back to the first . . .

When the Climate Changed

Evil's Enlightened Masquerade?

The kingdom came in the power of the cosmic Spirit, said Luke the writer of the third gospel. In the power of the Spirit, he continued, Mary of Nazareth conceived the king of the Jews. By the power of the Spirit, Jesus was baptized in the Jordan river and declared Yahweh's beloved Son. In the power of the Spirit, Jesus announced good news to the have-nots and freedom to the captives. In the power of the Spirit, Jesus had fellowship with the untouchables. And it was in the power of the Spirit that he healed. Eleven times we're told about Jesus healing children, women, lepers, and the sick.

But for Luke, the inbreaking of the kingdom didn't stop there: "With authority and power he commands the impure spirits and they come out!" (4:36); "Jesus cured many who had diseases, sicknesses and evil spirits . . ." (7:21); "he commanded the impure spirit to come out of the man" (8:29); "he was driving out a demon that was mute. When the demon left, the man who had been mute spoke, and the crowd was amazed" (11:14); "should not this woman . . . whom Satan has kept bound for eighteen long years, be set free on the Sabbath day from what bound her?" (13:16). Eight times in Luke's gospel we hear that the time was over for the power of demons.

In our scientific age, it's been common to downplay the existence of evil forces as one of the childish eccentricities of Christianity. Some of us who are engineers are tempted to write this off as ancient superstition. Prone to the flat and one-dimensional understanding of reality characteristic of our secular societies, we believe that we know better. We're convinced that the magnifying glasses of modern science actually make us better off, even if in countless ways they do.

Can they also blind us, however, from perceiving the unseen? Could it be that science studies that which we can measure or touch, but not what we cannot?

As a teenager, I listened to certain rock stars that opened me to the reality of evil. Watching certain bands on MTV, I remember experiencing a force that filled me with rage and prompted me to call out to a dark power, frequently. I also recall the story of an ordained priest who once visited my grandmother's house after she had experienced times of disquiet, only to find out that there were curses directed at her written on several voodoo dolls that someone had deliberately buried throughout her garden.

But beyond personal anecdotes, even a glimpse at history seems to prove that dark forces also operate on social and ecological levels. The Romans maintained their peace through subjugation by taxing millions with crushing fees and crucifying dissenters by the thousands. More recently, the Rwandan genocide or the German holocaust stand as proofs of what happens when entire populations bend their knee to worship their nation at the expense of "the others." It's enough, too, to look at the mountains of toxic trash currently dumped into China, or to the bleeding rivers in the Global South, or to the past forty years when the Earth has lost half of its vertebrate species as a one-sided flag continues to be waved in the name of consumerism and progress.

If this and more is not the work of a dark power, then it's hard to tell whose work it is. To paraphrase Voltaire, if the devil didn't exist, it would be necessary to invent it. Regardless, reasons like these led British theologian Lesslie Newbigin to suggest that in the name of science and of focusing only on what can be measured and touched, we have actually ignored the demonic to our peril.[146] Today we ridicule cosmic evil as if 'the devil' was a silly caricature with a red face, two small horns, and a pointy mustache. And yet in doing so, we ignore the darkness that always parades as light, and that the "father of lies" is seductive and often wears Prada (cf. John 8).

The Domination System Today

The question lingers: How are we to reflect, pray, and work toward the ecosystem of heaven as we face the threat of climate collapse? And how might we be called to *live out* Jesus' prayer for the coming of the kingdom in a day and age when "the domination system" (what the New Testament calls "the world") has artificially buffered us with the comfort of supermarkets and shopping malls?

Consider again a snapshot of the facts. For one, ecological economists point us to the fact that 1 percent of the world's people have greater wealth than 56 percent of the global population combined.[147] Others estimate that around 7 percent of the world's wealthiest emits more than 55 percent of greenhouse gas emissions while the poorest three billion release only 6 percent.[148] Picture this by realizing that two German Shepherd dogs in North America use more resources than the average person in Bangladesh does in one year.[149]

Such worldly consumer society has led sociologist Zygmunt Bauman to remark how most of us in the West have become "tourists and vagabonds," traveling and shopping our way mindlessly as we go around consuming and wasting away the blessings of the world.[150] Others suggest that just as the so-called German Christians supported Hitler after being blinded by the social fantasy story of nationalism, so today many Christians in the West have taken for granted and applauded the religion of consumerism.[151] We have turned greed into a virtue and shopping into a hobby.

Along these lines, ecologist and theologian Loren Wilkinson recounts the steady erosion of the living fabric that has made the territories of Canada the striking place that they are. For one, Wilkinson recalls how not long ago the powers-that-be repealed Canada's agreement to the Kyoto Protocol. But he also points out who the powers-that-be were involved in the removal of noncommercial fish habitats from the protection of the Fisheries Act, in the dissolution of the National Round Table on the Environment and Economy, in the termination of BC's oil spill response center, in the overstepping on First Nations communities by deciding to turn the Alberta tar sands into a continental carbon bomb, in the reduction of the number of protected lakes and rivers from 2.5 million to 159, and, not long ago, in the proposal of a federal bill to criminalize very much anyone who stands in the way.[152]

Power was meant to be exercised for good, but realities like these prove that the project goes off the mark—and sometimes painfully so. And the power abuse begs the question of the extent to which we are lured by fallen angels clothed in light to perpetuate today's global religion of endless extraction and comfortable consumption. Is it the case that the "domination system" has indeed lured us into the "mass psychosis" denounced by American biblical scholar Walter Wink?[153]

Cosmic Exorcism

Luke the gospel writer and John the seer might have thought so, for they testified to the reality that the earliest disciples saw and heard and touched as they encountered the ultimate power of the universe driving out what is not ultimate. In Jesus, they experienced the ultimate healer of the world coming to restore those who had been degraded, oppressed, or mistreated. But they also knew that the great come-down of heaven to Earth went beyond personal healings and liberations from evil. When Jesus was hailed as king as he entered Jerusalem on a donkey, when he overturned the tables and drove out the merchants who turned Yahweh's international house of prayer into a private national casino, when Jesus took on the fate of Calvary—the disciples knew there was more to it than met the eye.

A Climate of Desire

Perceiving the Invisible

Often we read these scenes as mere feats of journalism, as if Jesus was just an individual doing random acts in the society of his day and the evangelists were simply recording the random facts. In doing so, however, our individualistic blinkers often screen out the heart of the accomplishment, leading us to ignore the overtones of the drama. According to the penetrating vision given to John the seer as he wrote the Apocalypse, something far greater has happened. In poetic fashion transcending all sorts of this-leads-to-that rational logic, John depicted the victory of Jesus as unconventionally as one can, speaking of a powerful dragon that was defeated by the exalted Messiah:

> *The great dragon was hurled down—that ancient serpent called the devil, or Satan, who leads the whole world astray. . . .*
> *Then I heard a loud voice in heaven say:*
> *"Now have come the salvation and the power and the kingdom of our God, and the authority of his Messiah. For the accuser of our brothers and sisters, who accuses them before our God day and night, has been hurled down." (Rev 12:9-10)*

"A great dragon was thrown down . . ." Contrary to the caricatures, Walter Wink depicted demons as "the spirituality of systems and structures that have betrayed their divine vocations." These spiritual powers are rebellious invisible forces often clothed in light, luring and inhabiting people, institutions, and entire societies. And Satan, Wink said, is "the world-encompassing spirit of the Domination System"—the dark force *behind* all other evil forces—the unseen reality wanting to take over the cosmic show.[154]

Beasts and demons were the language the New Testament writers used to depict the inmaterial forces and myths that took hold of Jerusalem and the armies of Rome in the time of Jesus. Today's parallels are more subtle but no less real. Ronald McDonald could very well be a global corporate demon in disguise, damaging the health of millions with his hunger for a quick buck. The torch-bearing Lady may have very well been unwillingly turned into the visible window-dress for an invisible self-loving national angel, a federalist spirit inhabiting the most ecologically destructive empire planet Earth has ever known.

"But the dragon has been thrown down . . ." said John. The glitter of the "domination system" has lost its full grip. *Now* the salvation and the power and the kingdom belong to God and to the world's healing liberator. *Now* the clout is broken. *Now* the outcast who was silenced and despised reigns supreme above all names, all families, all corporations, all nations.

To be sure, John's was not the language of a polite, quietist man—even though politeness and piety are welcomed and important. His words cut through the pervasive imperial propaganda. His was and continues to be the full-hearted rhetoric of a Seer disclosing the invisible dimensions *behind* and *within* who Jesus was and what he accomplished.

> *For our struggle is not against flesh and blood, but against the rulers, against the authorities, against the powers of this dark world and against the spiritual forces of evil in the heavenly realms. (Eph 6:12)*

When the Messiah came, a war took place—not merely between Pilate against Jesus, and not merely between Rome and Nazareth. This was not merely an existential fight, not merely a nationalistic struggle. Far beyond, it was also a clash between the red dragon and the slain Lamb; between the domination systems of this world and the life-giving ecosystem of heaven. And heaven won. In their own subversive ways, the authors of the New Testament proclaimed the outbreak of a cleansing of the demonic forces that infested Yahweh's temple in Judea. They spoke about the outcasted king of the Jews performing the greatest of all exorcisms as he entered Jerusalem to endure the cross. They celebrated the beginnings of the cosmic victory of the kingdom of heaven over the kingdoms of the world. They bore witness to the triumph that came about because the love of the crucified God was stronger than the indifference or the hatred of the powers who crucified him. They testified to life wining and death losing, to the climate changing and the kingdom coming, to the earth beginning to receive the cosmic peace she had always longed for and desired.

Love Unleashed and Gone Public

What are some of the ripple-effects of the victory of the outcast? In what ways could we all respond accordingly in a world that is markedly different but altogether similar to the past?

There's certainly much to be said about faith communities embracing and fostering alternative economic arrangements. Confessing God as the living source of all things goes hand in hand with having a much greater care and appreciation for the things we have. It calls for bartering our way toward an economics of sharing and of gratitude and of true satisfaction. It implies beginning to pay the real price of our food out of respect for farmers and for the land and the watersheds. It means we can live on less and enjoy it more, recovering the longstanding spiritual practice of simplicity.

As sketched in chapter 5, there's also a great surge of inspiring energy coming globally from supporters of the fossil fuel divestment movement and steady-state, low-throughput economies. The energy transition and the pursuit of climate justice are nothing less than acts of public love. Debunking Wall Street and contributing towards a system change are badges of following the one who blessed those who work for peace (Matt 5).

And last but not least, the victory of Jesus summons us to walk from wronging to righting the greater victims of the world's fossil-fueled empires. Those of us who have benefited from the injustices of European colonialism should prioritize working toward reconciliation and companionship with First Nations and every other marginalized group. To them belong the stories that continue to be dwarfed and silenced by Babel's megaphones; theirs are the cultures that continue to be paved over by the highways of progress; they are the peoples who continue to be displaced from ancestral lands and who are already taking on the strongest brunt of climate change. Justice is due to them, and to all the generations that follow, aware that followers of the crucified Messiah have been promised inspiration from the Spirit to stand side by side and "mourn with those who mourn" (Rom 12:15).

Ecological Liberation

Upholding contributions like these, the overall witness of the biblical narratives may pose an additional challenge before us.

"We need a new Exodus . . . we need a mass peoples' movement." Such was the reply of a colleague of mine from A Rocha Canada when I asked him about the sort of response we need to face the climate challenge and the energy transition.

Like maple syrup had done so several years back, his words also lingered and stuck with me. And they stuck because there have been such movements in the recent past. Even as the kingdom of heaven often spreads subtly and silently here on earth, its reality is also public, energetic, concrete. For one, we could recall the ancient Israelites blowing horns and trumpets as they circled the falling walls of Jericho. But more recently we must also remember and take courage from the abolitionist campaigns in Britain, or the 250,000 daydreamers of the early civil rights movement mounting the steps of Lincoln's memorial to expel the national demons of hatred and racial segregation.

Is the time not ripe for all people of good will and for those who follow after Jesus to engage in redemptive acts of ecological liberation? Will we cry out and sing and work for freedom at the door of Egypt's palaces, reminding today's Pharaohs that only God is God and that they are not? Will we offer our whole selves and take part collectively in Spirit-led exorcisms of the glittering forces of death that are degrading the living world (cf. Rom 8; 12)?[155]

Such an ecological exorcism, of course, is not one of blind faith in techno fixes or green technologies (even though such technologies may have a role to play). It's also certainly not a matter of sticks and stones, or of anger, arrogance, or triumphalism. If anything, the two great wars and nuclear bombings of the last century remind us that we fall short of creating a utopia or bringing down the kingdom of heaven to Earth all by ourselves. Secularism may tempt us to worship our human powers, but the proven shortcomings of that faith should always remind us of the importance of God healing our inner darkness. Calling for transcending the narrow vision of our secular world, perhaps it was reasons like those that led a former president of the Czech Republic to remark that:

> The Declaration of Independence states that the Creator gave humans the right to liberty, but it seems we can realize that liberty only if we do not forget the One who endowed us with it.[156]

The enormous work ahead calls us instead to work in healing companionship with the cosmic Spirit, inspired and indwelt by Yahweh's breath. Our shared struggle is one to be engaged in faith and hope, in joy and grief. And, above all, it's a struggle of love, of fighting lies and mocking false gods, of giving away our bodies for the life of others.

Aware of such struggle, Old Testament scholar Christopher Wright calls for a "spiritual warfare" that is marked by:

> deep compassion for those oppressed by the forces of evil and idolatry—with all their attendant social, economic, political, spiritual and personal effects. We battle with idolatry because, like God whose mission we thereby share, we know that in doing so we seek the best interest of those we are called to serve in his name. We combat idolatry not only to glorify God but also to bless humanity. Spiritual warfare, like all forms of biblical mission, is

to be motivated and exercised with profound love, humility, and compassion—as modeled in Jesus himself.[157]

Imagine for a minute what it might imply to put fasting and public prayer, poems and songs of liberation, at the service of this cosmic gospel. Imagine the word of Yahweh not returning empty but spreading unbound, accomplishing that for which it was sent (Isa 55). What would it look like to participate with the living God in an ecological exorcism that heals the living world? Is there hope for the struggle?

A Past Filled with Hope

Perhaps more then than today, slavery has been one of the most entrenched practices of human civilization, and yet for some, "normal" did not mean "normative." Less than two centuries ago, a committed group of followers of Jesus responded to the call of the prophets to break the bonds of oppression. They wanted a fair life for the women and children torn away from their families to work in distant mines and plantations. They found it hard to accept to throw more than 250,000 corpses of dead African slaves into the unruly waters of the Atlantic. But knowingly or unknowingly, these folk were grasped by the conviction that the love of God was greater than the bestial forces of economic profit. Christian evangelical William Wilberforce and the abolitionists saw no obstacle in the twenty million pounds it cost the British to set their slaves free in 1833. So time after time, they set themselves to change the rules of the game, and, even if imperfectly, they did.[158]

Like Abraham and Sarah and all those who have followed after their steps, many more have gone down in history as practitioners of this way. To mind come, of course, the courage of Rosa Parks and of the great Martin Luther King Jr., who embraced the divine dream of love over the nightmare of violence and the numbness of resignation. Needless to say, both of these leaders in the movement for civil rights had deep roots in the rich soils of African-American Christianity; Parks as a lifetime member of the African Methodist Episcopal Church, and King as a once half-hearted skeptic that went on to become a baptist minister whose flare and non-violent work remain unmatched to this day.

In Canada, recall again the social legacy of Tommy Douglas, who left his baptist pulpit to give himself to the ministry of public love and become one of the architects of the nation's healthcare system. Or remember Jean

Vanier, whose influence is still felt; be it though his moving humanitarian writings or through the L'Arche communities he helped found, now providing over 145 homes of belonging and support for folk with intellectual disabilities across five continents. Likewise, on a much smaller scale, I'd like to recall my former church community in Vancouver, who recently raised $9 million toward a social-housing complex that now provides a place of belonging and affordable accommodation for all sorts of people across the economic spectrum.

In my native Costa Rica, too, this dream of public love has also found multiple expressions. One of them came into being in the 1940s, as clergyman monseñor Sanabria joined many who wanted to respond to the disequilibrium caused by the increasing concentration of wealth and power. Despite the pressures from some ruling elites, Sanabria and his maladjusted fellows persisted nevertheless, working together with labour unions, well-meaning common folk, and other political figures, many of whom had a social-Christian bent. Defeating convention while drawing from the profound social vision outlined in *Rerum Novarum*, Sanabria's leadership in the Catholic Church played a foundational role in changing the country's constitution to include labor codes, diverse social insurances, universal healthcare, and, last but never least, paid vacations.

But men like these were not alone. Whether as mystics, as mothers, as missionaries, or as martyrs, women following after Jesus have also left a mark in Christianity. It was in fact mostly women who cared for Jesus during his life, death, and burial; it was Mary Magdalene who eventually became the first preacher of the good news; it was Prisca, Juna, Chloe, Lydia, Nympha, and Phoebe who are recognized in the New Testament for their service as prophets, church leaders, and apostles.

Following their steps, to mind comes a third-century young member of the noble class Vibia Perpetua. This woman's fearless faith eventually led her to defy some of her biological father's expectations and join, instead, her new family in Christ. In doing so, not only did she set an early example that flipped the Roman father-centered mold upside-down. She also proved that women could have a direct relationship with God in a male-dominated culture that gave that authority only to churchmen. Likewise, Perpetua's life also challenged the inequality that pervaded Rome's empire, especially when she was executed alongside other slaves for confessing themselves as being Christian. Her life and death became a dramatic statement about Christianity's ability to transcend social distinctions, while being an expression of feminine courage in the cosmic battle against the Domination System.

Less visible and certainly more subtle were the footprints of Monica, a fourth-century woman born in Thagaste. Not only was she biological parent of someone who came to be one the most formative thinkers in the West, Augustine of Hippo. She was very much his spiritual mother too. According to Augustine's testimony in his *Confessions*, it was Monica's unstopping devotion to Christ that served to bring him to spiritual birth, and it was in her voice that Augustine came to hear the voice of God. Not only did Monica nurture the life of this one theologian whose writings, for better or for worse, would come to have a massive influence on Martin Luther, John Calvin, Jonathan Edwards, and the many generations of Protestants that have followed. Without knowing it, Monica also contributed indirectly to shape the way we all see ourselves today as unique persons of infinite worth. To be sure, this is a belief we've inherited mostly from her son, who came to appreciate himself as supremely worthy because of God's personal love for him. But this Augustine learned not only from the poetry of the Psalms; he came to know that first-hand because of the divine love that he himself acknowledged to have experienced in a special way through his mother. Regardless, little did he foresee that this North African woman would become one of the greatest spiritual grandmas of Western culture.

Despite being silenced, ignored, or brushed away in what was a world controlled by men, next to Perpetua and Monica one must add Blondina, Columba, Clotilde, Dhuoda, Ethelburga, Erentrude, Eustochium, Felicitas, Helena, Hilda, Hroswitha, Hugeberc, Irmengarde, Loeba, Macrina, Melania, Marcella, Marcellina, Paula, Pope Joan, Radegund, Tetta, Theodora, or Walburga. Women like them were apostles, deacons, martyrs, widows, virgins, abbesses, monastics and missionaries whose testimonies of persistent courage and innovation survived the first thousand years of the history of the church.

Beyond individual people, much, too, could be said about the social achievements of entire Christian communities in the past. True, there were the crusades, and true, some Christians have blindly held the Bible in one hand and the flag of empire on the other. But, for better or for worse, it was under the influence of Christianity over more than a thousand years that European societies gradually recognized the worth of seniors and infants. It was largely because of Christianity that the Roman gladiatorial games were abolished and that the grounds were laid for the proliferation of orphanages and public hospitals and "houses of lepers." It was Christian monks who established medieval universities throughout Europe and who legislated to constrain patriarchy. It was a countless number of unsung heroes and heroines following after Jesus who helped to infuse Roman justice with mercy, who upheld the value of women, who made room for the poor in law courts, and who turned their homes into places of welcome for the suffering and the outcast.[159]

As this history unfolded with all its ups and downs, even main proponents of secular humanism like Richard Rorty acknowledge the historical importance of the Bible in laying down the foundations of today's world. It was the controversial Martin Luther and other Protestant reformers who stressed the importance of public education not only for men but also for women and children. It was the Franciscans that sowed the seeds of what today we know as micro loans to the poor. It was Christian reform groups like the Jansenists who defended the right to oppose the tyranny of kings and corrupt churchmen, and that provided crucial religious ideals from early on that shaped what ended up being an (anti-Christian!) French Revolution.

Surely, this is not to say that the rule of the majority is without problems, cause it clearly ain't (it was the voting booth that initially got Hitler into power). But while in the West we have come to take democracy and universal human rights for granted, it should at least remind us that their vision of equality and justice for all was grounded in the precious worth that *all* human beings have to the God revealed by Jesus, as Christianity has always affirmed.[160]

∼

A sketch like this could be dismissed as simplistic and somewhat romantic. One could also affirm that doing good is not the monopoly of those who follow the steps of Jesus (it's certainly not), or that Christianity was a random accident of history and that now we should forget its roots and simply enjoy its positive legacy and enduring fruit.

Fair enough. Cypress or eucalyptus branches that sit on a vase will continue to spread their fresh scent long after the water runs dry. And it is true, as well, that we church folk often have a checkered history—with the evils that the church perpetuated in the residential schools of Canada being a painful reminder of that. Sometimes the fans of Christianity have indeed been Christ's worst enemy.

But the gospel cannot be sold short. None of the achievements above trickled down like bread from heaven in the bloody arenas of the Roman Empire. It was not in the sanctuaries of Aphrodite or Zeus that entire cultures were gradually infused with a spirit of service, compassion, and benevolence for all. It was not the gods of Babel, Egypt, or Rome who outpoured themselves for all humanity and for the healing of the cosmos. This and more was the bumpy, imperfect work of the people of the God and *Abba* of Jesus of Nazareth—the God of justice and love who made the best of three iron nails and one rugged cross to turn around the climate of history.

And turn around it has, for now this great cloud of witnesses continues to sing out loud reminding us that today is the day for all of us to join them and collaborate with the cosmic Spirit in the healing of the world. The angels sing and the demons tremble when the living presence of Yahweh goes public.

> *The seventy-two returned with joy and said, "Lord, even the demons submit to us in your name."*
>
> *He replied, "I saw Satan fall like lightning from heaven. I have given you authority to trample on snakes and scorpions and to overcome all the power of the enemy; nothing will harm you." (Luke 10:17-19)*

∼

As we look at the challenging times ahead of us, it is hard to tell if we will get our act together to keep the climate from changing even more drastically. But we know that God can because, in the past, God has done so. So the question is not so much whether the living source of the cosmos could unleash a new collective miracle, of a sort that we are yet to witness. The question is altogether different: Will we join in? Will *you* join in?

7

Becoming Earthkeepers
A Call for All of Us

"Here we don't have watches but we always have time."
—African saying

"I don't come to solve anything.
I came here to sing
And for you to sing with me."
—Pablo Neruda

AFTER THE CRYSTAL DOME was cracked open, the skies lit up again, and the people were set free. That's arguably one of the most illuminating scenes of the 2015 film adaptation of Antoine Saint-Exupéry's *Le Petit Prince*. The movie tells the story of a schoolgirl who visits different planets alongside a witty fox and a peculiar prince who often poses odd questions to the characters he encounters.

At one point in their journey, the action slows down on a planet that a businessman had claimed as his own. But owning the planet was seemingly not enough for him; the executive claimed the stars as his sole possession too. Busy as he was accounting for them, it turns out, he likely became blind to the fact that his hunger for profit came at the expense of turning the entire planet into a clunky gigantic factory and a monochromatic office complex. In fact, there were hardly any lights in the skies because the businessman had gone as far as to capture the stars and then squeeze them into a gigantic crystal dome of sorts. Aided by who knows how many technocrats, he then managed to suck their energy and use it as the power source for the planet's offices and factories.

Surviving the mild but perpetual gloom, folk didn't smile much. Living in the businessman's planet meant joining the treadmill: people looked downward when walking and downward while working. Rather rapidly, everyone got used to the seemingly inevitable exchange. A concrete jungle with fluorescent light bulbs on the inside meant a dull routine, gloomy night skies, and piles of trash on the outside. Efficiency and profit trumped splendor and wonder.

In response, the prince, the fox, and the schoolgirl kicked off a rather daring rescue operation. Having freed their clunky aviator plane from the three-pronged claw that had trapped it, after a few seconds of flight the girl jumped courageously to land on top of the crystal dome. She then went on to shove a sword into it, opening a crack in the power source in what turned out to be a decisive turning point of the story.

Gradually squeezing themselves out, the stars began to rise like angelic beings finding their way up into the infinite skies where they once belonged. Not long after, the power plant stopped working, light bulbs dimmed out, the city came close to a full stop. Certainly to the businessman's impatience, everyone began to leave their desks behind, for both young and old equally started to marvel at the forgotten magic they experienced outside. The tradeoff was clear: as the evening skies recovered their radiance, people began to recover their freedom.

∼

Starlight also features in a sacred story from the Book of Genesis. Following the scenes of Yahweh's everlasting commitment to "all living creatures of every kind" and to "all life on the earth" (Gen 9:16–17), we are then invited into one of the unique scenes in the Hebrew Bible recounting a special revelation given to Abraham:

> [Yahweh] took him outside and said, "Look up at the sky and count the stars—if indeed you can count them." Then he said to him, "So shall your offspring be." Abram believed Yahweh, and he credited it to him as righteousness. (15:5)

"Look up at the sky and count the stars, if indeed you can count them." The impossibility of the task was obvious: the stars are innumerable. But Abraham didn't brush it away, either by dismissing the call or by flattening the experience by taking a quick snapshot with his iPhone. Similar to the people who marveled at the businessman's planet, Abraham also came to stretch his sight beyond the tip of his nose.

"Look up at the sky"; "count the stars." The Hebrew verb *habbet* (הַבֶּט) implied taking a long, reflective look. The call required attentiveness to the infinite constellations. Like that of the psalmist, it's likely that Abraham's faith grew stronger by recognizing his small stature in a cosmos that he could never fully grasp or understand. Lost in wonder, perhaps the vastness of the open skies stirred him open to a reality far beyond himself—in this case, all the way to the point of confirming his trust in Yahweh. It's hard to know. But what is clear is that Abraham's was no inward journey; his faith grew stronger when he heeded the call to become attentive to the cosmic glitter waiting for him . . . outside.

From Neon Light to Starlight

To be sure, the film adaptation of *Le Petit Prince* is at times marked by caricature and exaggeration. Some factory owners are not greedy or insensible. The narrator of Genesis tells us too that Abraham and Sarah's faith was faulty and riddled with compromise throughout their life. What is more, many of us who grew up influenced by the short-sighted pragmatism of the modern scientific mindset will likely dismiss these stories as childish nonsense.

But the dismissal might not serve us well, especially if we ignore Einstein's reminder that we can't solve problems by applying the same logic that created them. As we seek to respond to our ecological emergency, green techno fixes and short-lived political band-aids will not be enough. Behaving as if the world is a disposable amusement park will not cut it either. Eternity is waiting for us to leave behind the indifference, the desire for quick profit, and the technological domination of nature that we continue to exercise in our me-centered shopping world. Here one could dare to paraphrase Jeremiah and admit that ecological intercourse without friendship and commitment borders on rape.

Becoming *earthkeepers* and bearing witness to the sacred intercourse the world is destined for is a call to go far beyond. It means living into the recognition that we are one among countless members sharing in Earth's living community, our only home. Earthkeeping calls for a noble way of going about being human, summoning us to leave behind the modern skepticism toward the supernatural while embracing empathy and attentiveness as postures needed to heal our relationships toward the living world and between each other. Not least, the journey ahead calls for experiences of faith and marvel as well, like those of Abraham's starry night or those of the liberated citizens of the formerly gloomy planet in *Le Petit Prince*. Becoming earthkeepers means stretching our aspirations into the infinite horizons of eternity, recognizing that the furnaces of hope will light up once we allow the stars to shine brighter than our neon lights. The sadness that covers creation will begin to turn into joy as the glimmer of our screens is dwarfed by the heavenly radiance of cosmic starlight.

What might it look like to live as guests in a world that is not of our own making? What will revive the sparks of change as we undertake the conversion that German theologian Jürgen Moltmann called for in moving away from lordship to fellowship, from conquest to participation, from production to receptivity?[161]

I'd like to conclude by sharing a few reflections aiming to honor the Hebrew people of old who recognized this sacred planet we share with countless creatures is destined to become nothing less than Yahweh's temple.

Healing for Our Companies & Our Cities

Beyond Nature as Machine and Life as an Equation

Unmasking the effects of dehumanizing technologies in his *Psychology of Science*, Abraham Maslow once remarked that if the only tool one has is a hammer, then one treats everything like a nail. Likely not something most of us were taught in business school. In fact, many of today's standard business practices continue to be informed by standard management theories. Robert Kaplan and David Norton encouraged us to think of organizations as airplanes and of balanced scorecards as the instrument panels in the cockpit. Harvard's Michael Porter promoted his "shared value" framework as the new "key" to "unlock" and "drive" business growth.[162] The revolutionary management philosophy popularized by Eliyahu Goldratt in his 1984 classic, *The Goal*, argued that adopting "making money" as the goal of a manufacturing organization looked like a pretty good assumption.[163] Goldratt saw productive processes as an uneven chain requiring optimization.

These are all examples of how the metaphors we use often betray us—and by doing so continue to hold us captive without our awareness or consent. The nature-as-machine view of the world that has shaped society's institutions is largely a result of 250 years of cheers and praises to the Industrial Revolution. In our increasingly urban contexts, our imaginations have grown accustomed to seeing the world through mechanical (and now virtual) lenses. We speak of the economy as an inert reality and of business as a machine steered in a profit-hungry cockpit. We are led to diminishing ourselves by believing that Wall Street, Hollywood, and the Pentagon call the final shots, that the environment is something "out there," external to us, that we're simply human "resources" in an increasingly disposable labor market, and so on. (Or, so, at least, have King Coal and Mr. Profit allured us into thinking.)

Yet, if our ecological track record were to speak for itself, then the one-sided flag of consumerism and progress would prove them wrong. These and other slogans of empire continue to be a dishonor to the God of life. And they beg the question of whether economics and technology should continue to be our masters, and of whether minerals and fossil fuels should be endlessly extracted to keep turning a living planet Earth into a hyped-up version of George Lucas's Death Star.

Liberating the Imagination

There are sprouting signs of hope beyond our technological love affairs. Likely unaware of the earthkeeping vision of the ancient Israelites, and despite their ongoing ties to the consumerist paradigm, progressive companies and cities are nevertheless taking bold steps toward truly circular business models with a long-term vision.

Interface's "Mission Zero" is arguably the best-known example of a company taking an audacious move toward drastically reducing fossil fuel emissions. It's also becoming a zero-waste, 100 percent recycled-fiber carpet enterprise, where the output of one cycle becomes the input of another.

CEO Ray Anderson experienced important realizations along the journey of healing his petroleum-intensive company until it became the world's model manufacturer in today's carpet industry. To begin with,

Anderson admitted that he had been "plundering" the Earth and his children's future (a necessary confession often bypassed even by some of the best-intentioned sustainability practitioners). But most revealingly, the superior heights reached with Interface's magic carpets did not come about by consulting a technological genie in a plastic bottle. Reorienting his company to reach the summit of what he called "Mount Sustainability" eventually translated into flooring solutions modeled on the way the natural world works. "We don't believe anybody (ourselves included) can make a green product in a brown company." The shift away from a linear take-make-waste model of production toward a cyclical and restorative one, was grounded in a desire to respect and learn somethings from nature's patterns.[164]

The living world has also shaped the vision and practices of outdoor clothing company Patagonia. When asked to compare his company's priorities with the imperative of endless growth in today's business world, CEO Yvon Chouinard pointed to how Patagonia is modeled on the slow, gradual rhythm of trees growing in a forest until they stabilize. Not too fast, not too much, but only one ring at the time.[165]

Is that not sacred wisdom for our accelerated age? Their grandeur and splendor aside, forests breathe in greenhouse gases to release oxygen. They filter and recycle water and protect soils from erosion while doing so. Forests grow at a gradual pace, taking in only what they need, even as they provide a living habitat for countless other creatures. They live and let live. Not surprisingly, trees have existed sustainably for thousands of years thanks to their equitable trading practices that allow them to grow into maturity and then reach relatively similar heights. Perhaps Ecuadorians put it best in recognizing that forests don't need development because forests *are* developed.

To this, some will be quick to object that lions or sharks could also serve as natural metaphors. In fact, standard management theories selectively praise these predators by highlighting them as exemplary creatures that hunt voraciously after their prey. But in doing so, such theories ignore that lions and sharks kill for need, not for greed—and certainly not for sport.

Can business prioritize need instead of greed? Uncommon as it is to imagine restaurants closing their doors on Sundays, at least one chain in the United States is doing so. Respecting the biblical tradition of keeping Sunday as a common day of rest, Chick-fil-A has kept its restaurants closed on that day ever since the company's foundation in 1947. While the politics of Rome and Babylon continue to brainwash us into believing that humankind was made for business and that business was made for profit, Chick-fil-A has exemplified a contrasting vision that recognizes that business exists for the service of humankind and that we humans live as guests in a much larger world that's not our own.[166]

Municipalities are also responding to callings which are higher (and sometimes contrary) than those of federal governments. Despite the US presidency and its climate-change-denying cabinet pulling backward, more than 400 mayors of US cities are smartly working forward to reduce their fossil fuel emissions. In doing so, they have in fact joined over 7,100 cities around the world committed to the cause, which is no small feat considering cities themselves have the potential to reduce up to 80 percent of the global GHGs.[167]

Citizen groups, as well, are showing promising signs of embracing the renewable way. In my native homeland, Costa Rica Limpia is advancing long-term, citizen-based coalitions for the country to build an entire transportation system based on renewables. No small task, but inspired by the nation's legacy of abolishing its army, Costa Rica Limpia is now set to abolish the country's dependence on fossil fuels.

Similarly, in Québec, Canada, numerous indigenous groups, labor unions, professional associations, and NGOs played a decisive role in delaying, and finally stopping, the Energy East oil pipeline. Lawsuits, peaceful direct actions, speaking tours and artistic skits were some methods which were deployed in order to stand up to short-term corporate interest. As elsewhere around the world, they proved that federal governments did not have the last word, and that flat financial calculations cannot contain the groundswell of committed citizens and land-protectors safeguarding life and prosperity.[168]

Faith communities as well have been walkingliving into ecologically grounded paradigms. For one, more than 210 faith organizations and congregations have joined more than eight states, 1,780 businesses and investors, and 335 colleges and universities demonstrating their support to help the US achieve its promised GHG reductions under the Paris Agreement.[169] Going further, many others are living into what American theologian Ched Myers and his colleagues are championing as "watershed discipleship." Recognizing that the land and waterways precede us and have an integrity of their own, watershed discipleship honors and *works within* the particular physical contours of a place. It goes beyond the placelessness of more abstract efforts of creation care (which often end up in simply changing light bulbs and recycling) and instead acknowledges that the earth's bodily topography overrides artificial, human-made political boundaries. To be a disciple of one's watershed means taking a stand on the immediately proximate watershed context each human community inhabits. People in the town of Fort McMurray, Alberta, for example, can do little to keep Venezuela from extracting oil, but they can do much more in actively resisting the pollution of the very river running through the Athabasca watershed. Like earthkeeping, watershed discipleship recognizes that we are not gods, that we cannot do it all. But precisely *because* we're small, precisely because we're human, made from the very dust to which we will return, it does call us to be replaced and regrounded together in each of the millions of unique ecological cradles that sustain us.[170]

Countless case stories like these prove that earthkeeping is not a simplistic protest to go back to the bush. We simply cannot turn the clock back and suddenly do away with our 10,000-year-old civilizational experiment. The progress we've made, as Ronald Wright remarked, has actually destroyed the lower rungs of history's ladder and we've reached the point of no return. We must take off from where we are.

Earthkeeping is also not an adoration of the natural world, as if it were divine and should be revered and left untouched. Nature itself is bound to patterns of violence, darkness, and death—patterns which the writers of the New Testament declared to have been mysteriously undertaken and transcended in the ministry, death, and resurrection of Jesus.

The liberation that we need calls, instead, for nature and culture to coexist in reciprocity on this side of eternity. It calls for us be content with what we have instead of craving endlessly for what we don't. It calls us also to treat our living home as a sacred gift, recognizing that our long-term stay will depend on us ceasing to behave as masters and start living as guests, treading lightly in a world that was there long before us and that is certainly not of our own making.

Doing Good by Overcoming Evil

Taking positive steps toward a low-carbon future calls as well for overcoming society's habits, desires, and systems that stand against sustainability. It would otherwise be like asking a person to love his spouse but remaining silent about his relationship with his lover, or like cleaning waters downstream but doing nothing about their pollution upstream.

Deeper change is always two-sided. Walking away from consumerism and indifference toward sustainability and *earthkeeping* means living in a way that bears witness to the overflowing abundance of the ecosystem of heaven. It also calls for deflating what Ronald Wright despised as the "consumerist pornography" of advertisement as a way to resist the military-industrial conspiracy to murder creation grieved by Wendell Berry. Saying "go" with one hand requires that we say "stop" with the other—for love is fierce and passionate, and it holds fast to what is good but abhors what is evil (Rom 12:9).

Revisiting God and Caesar (and Corporations)

In the letter to the faithful in Rome, the apostle Paul urged believers to "be subject to the governing authorities, for there is no authority except that which God has established . . . For rulers hold no terror for those who do right, but for those who do wrong" (Rom 13:1–3).

For better or for worse, experience and necessity show that governing authorities have a role in society. But in directing such words to a minority of believers living under the shadow of Rome's empire, did Paul endorse free game for the powerful to set the rules and blind trust by the rest to obey them?

The statement has often been tragically bent out of shape. As was the case by some in Nazi Germany who appealed to Romans 13 in their support of Hitler, some Christians still believe Paul's words are a blank check endorsing the legitimacy of *any* political regime. But a closer look reveals how subversive the statement actually was. For one, the fact that someone comes into a position of authority does not mean he or she will actually exercise that authority justly. King David became a murderer, King Solomon, a womanizer. More importantly, Paul implied that God-honoring rulers are *not* to hold terror on those who do right. Authorities are called to be servants for *good* (13:4). He subversively reminded believers in the capital of the empire (where Caesar lived) that there is "no governing authority" which is not *under* the authority of God. In other words, they might claim their coins as their own, but *no religious, political or corporate leader has absolute status or final word.*[171]

This is surely no endorsement for the church to become the moral police of the universe. It's also not a call for mindless anarchy, nor a propped-up political slogan affirming that democracy by itself will bring about the bright day of history foreseen by the biblical authors. Not unlike the rest of the prophetic writings in scripture, Paul words remain a clever and oblique indication that leaders can and must be regularly called into account when they fail to exercise their authority humbly and justly.

Ask John the seer. In light of the arrogant outrage of beastly powers set up against the living source, John's apocalyptic visions also shed colorful light on imperial politics, then and now. Symbolized by the lampstands and the sackcloth, John portrayed faithful churches as those who grieved the presence of evil powers, but who also spoke up to them. Their words were like a fire that burned through Rome's glittering propaganda, testifying instead to the Lamb seated on the throne of the universe. "Worshiping the creator and redeemer as the sovereign one who rules," says historian N. T. Wright "was and is fighting talk in the world of Caesar's empire."[172]

Surely, this is fighting talk prompted by compassion and not by hatred. Neither Paul nor John called for wild anger, religious pride, or violent revolution. They knew very well that living under the shadow of beastly powers required enduring patiently and overcoming evil by doing good. But by no means were they endorsing shrinking back from the world that Caesar had arrogantly claimed as his own. Instead, they called faith communities to *engage it* by bearing witness to the light—a challenge just as alive today as it was back then.

Today, the quest for climate justice is hijacked by laws, corporate interests, and crafty governments that are often short-sighted and set up against the very integrity of the living world. And the hijacking begs the question: If they overstep their limits, should the powers-that-be be ignored? Prayed for? Confronted?

In her eye-opening *Resisting Structural Evil: Love as Ecological-Economic Vocation*, American theologian Cynthia Moe-Lobeda proposes six "key gateways" to effect deep-seated system change in a way that protects workers, communities, and ecosystems. The gateways include business alternatives, changes in moral cultures within institutions, and citizens reclaiming governments to restrain and redirect corporations. Confident that faith can move even corporate mountains, the book ends with practical guidance for churches and citizen groups wanting to give allegiance to the living source of the cosmos far beyond giving it to short-sighted corporations or self-obsessed patriotic governments. Here one cannot restate what Cynthia has eloquently said already, but simply encourage more curious readers to engage with her final chapter, where she expands these six gateways as a way to stop self-serving governments and corporations from having the ultimate word.[173]

The Way of Peacemaking

Adding to a manifesto like hers, Martin Luther King Jr.'s sobering 1963 "Letter from a Birmingham Jail" sheds light on our challenges at hand. Reflecting on the many times he was arrested and mocked for the cause of civil rights, King urged his fellow churchpeople to grapple with important realities regarding the confrontation of an unjust law—which, agreeing with St. Augustine, he recognized to be no law at all. King went on to recall Hitler as someone who acted legally, contrasting him with the three biblical characters in the Book of Daniel who were thrown in jail for refusing to obey the self-serving laws of King Nebuchadnezzar of Babylon. He also recalled the early Christians who faced the lions for resisting to submit to the unjust decrees of the Roman Empire.[174]

Inspired by Jesus' taking of the cross to expose and disarm the shameful ridicule of its violence, King called churches to embrace the consequences that came with breaking unjust laws as the clearest way of exposing their very injustice. "One who breaks an unjust law," he said, "must do so openly, lovingly . . . and with a willingness to accept the penalty." In turn, he recognized that the person who accepts an unjust penalty to "arouse the conscience of the community over its injustice, is in reality expressing the very highest respect for law."[175]

As Dr. King knew much better than any of us ever has, peacemaking resistance is no glamorous challenge. For one, the task is humanly unsettling and likely terrifying for most of us if left on our own. It also calls all who are engaged in the climate challenge to work together across differences of culture and personality. And it calls us, as well, to overcome our drive toward autonomy and to seek, instead, inspiration and strength from above as we stand together alongside the folk facing the greatest consequences.

Babylon's social pyramids elevated the king as the sole person representing the chief god. Regardless of age, gender, race, or religion, the Book of Genesis protested in contrast: *all* human beings bear the image of Yahweh, the one true God. Far from being an imperial despot, Genesis speaks of Yahweh going into the risky business of delegating a great deal of authority to all humans in a democratized kingship of sorts. That's why different expressions of Christianity remind us how the God Jesus revealed are often found in the faces and the stories of those who are forsaken. Cutting through today's smokescreens of entertainment, we discover that the nobodies of societies are often sources of divine encounters—be it displaced shepherds in the rugged hills of Peru or the plains of Nicaragua, or migrant workers in the dry fields of Andalucía or California, or women and children facing floods near the Padma River in Bangladesh or along the long banks of the Mississippi.

We also discover that the grounds for listening have been laid down for our generation. Just a few decades ago, many white people joined African-Americans in their journey toward civil rights. Today many European descendants are walking the walk toward peace and reconciliation, upholding the infinite value of land and watersheds, even as they seek to respond to the historical wrongs done to aboriginal populations.[176]

I was privileged to have a glimpse of this journey during the years I lived in Vancouver, the territory traditionally inhabited by the First Nations of the Salish Sea. I remember the mayor of Vancouver joining the global climate march while using a megaphone to urge the Canadian Prime Minister to keep his climate promises. I recall an Anglo-Saxon instructor of mine who eventually stepped down from his professorship to give himself to a social justice action group along with his wife, seeking to live into the biblical prophetic tradition. I came to admire a good Canadian friend who was detained by the federal police as he peacefully stood alongside ignored aboriginal peoples that said "no" to the expansion of the Kinder Morgan oil pipeline, but "yes" to clean waters and to a stable atmosphere. I remember a Caucasian woman who visited some aboriginal land-defenders sitting peacefully in the way of the proposed pipeline expansion, but approaching them without saying "I'm here to fix your problem." She simply asked "What do you need? How would you like me to help you?"

Becoming Earthkeepers

Costa Rica is a small country that ever since colonial times has looked up to rich nations. But being the Costa Rican, the *tico*, that I am, folk like these give me hope. For one, they give me reasons to recover a sense of agency, having been raised to believe that the keys to history belonged to the superpowers. But they have also helped me realize that money, race, fear, and privilege don't have the last word because there is a living source of sisterhood and brotherhood much greater than any of us can fathom or imagine.

Rediscovering Forgotten Horizons

Earthkeeping acknowledges that the atmosphere interweaves us all, human and nonhuman, and that caring for it is caring for all. It also recognizes that we have been wounded by darkness; that in one way or another we all carry an inner violence within us. And so it faces us with the question of whether we can heal air, land, and water, without being healed ourselves.

Modern science has come to categorize us as *Homo sapiens*—human beings with the capacity to be intelligent, reasonable, and sometimes wise. There is obviously truth to that. We build space telescopes, we compose songs, we manage to do heart transplants, we write policies and laws, we design mobile phones and their ingenious apps. These and countless other achievements are the remarkable results of building our Western world on the belief that more education and more knowledge will improve our societies. To a great extent, they have and they will (one doesn't write books if one didn't believe so).

Still, for a good 200–300 years we've largely pushed the world of spirit to the side in the name of material progress and secular education. Contrary to common experience and to the poetry of creaturely compassion found in Psalm 104, we've treated seals and whales (let alone the cows and pigs in our feedlots) as imperfect creatures without feeling. We've assumed that anything that is not human is inert stuff that has value insofar as it can be turned into merchandise for quick profit. We've seen ourselves as the ultimate stage of evolution.

So there is little question that our minds need to be reenlightened and our imaginations liberated. Living under the looming shadow of climate collapse, the time is ripe for us to infuse every discipline with deep respect and gratitude for life.

A Climate of Desire

Beyond Books, Beyond Screens, Beyond Reason

Already around the turn of the fourth century CE, Augustine of Hippo recognized that our truest essence is being beings of adoration. We are shaped, not so much by what we *know*, but by what we *love*—and, in fact, by *who* we love. We are bundles of desire. So, embracing the ecological conversion required of us leads us to realize that perhaps a greater challenge beyond *thinking* differently about our place on Earth is that of *feeling* rightly. It might just be that, as we transcend our culture of instant hook-ups and quick fixes, we need to awaken a deeper sentiment of affiliation with God, with our fellow humans, and with other forms of life. Beyond enlightening the mind, we need our living source to heal and educate our sentiments. More than ever before, today we ought to be *moved* into the vast seas of empathy for all, far and near, human and nonhuman. We need to allow ourselves to be *pierced* by the untamable beauty and grandeur of the earthly temple we've been invited to care for and inhabit.

This journey might not be all too different from what the evangelists said Jesus experienced when he saw those who were harrassed and helpless (Matt 9:36). Being largely a man of the streets and the outdoors, Jesus was someone alive to the reality around him, and as such we're told he was "*splangkhnizomai*" (σπλαγχνίζομαι)—literally, "moved to his bowels," which the ancients recognized as the seat of love and pity. Not that there is anything inherently wrong with analyzing stats in a corporate room, or with being a clicktivist on Facebook, but as far as this story is concerned, the moving of bowels comes from experiences, which lead to emotions, which lead to actions. So in an age of snapchats and fast scrolling, perhaps much more than our minds, it's our guts, that need awakening and healing. Despite the tempting glitter of our screens and neon signs, life is waiting for us *outside*, calling us to take wonder in the cosmic light feast emanating from the stars.

I write all this speaking to myself, sitting in front of my laptop on a spacious wooden desk. As my mind travels at 100 miles per hour (even as my body is hardly moving), once again I realize the need for me to feel. Like most of us, I'm also becoming a prisoner of my screens; slowly forgetting to engage with others and with the land and waterways around me. Useful as it is, the virtual world is a mind-game of 1's and 0's. It's gnostic. The internet offers information and knowledge, but it fails to engage our whole selves. Along with Ken Wilbur, we can admit that most of us have not yet lost our minds, but that we're close to losing our bodies—and the planet that nourishes them.

One of the artificial alchemists of our modern mind maps, René Descartes, made us believe that the mind and the body were separate things. Another, Emmanuel Kant, was convinced that it was our supreme ability to think that made us uniquely human. But the ancient wisdom of the descendants of the Hebrews calls us to reconsider. According to the poetry of the Book of Genesis, we're made for the earth, for each other, and for God; we're a sacred blend of Yahweh's breath and earthly stardust (Gen 2:7).[177]

Just as one learns to swim by jumping into the water (and not by watching a movie or reading a book), one learns to feel by feeling. Following theologian Steven Bauma-Prodiguer, *earthkeeping* recognizes that we can most deeply care for that which we know and that we most deeply get to know that which we are able to experience. More than ever before, the time is ripe for overcoming our "nature deficit." Engaging in the sacred intercourse for which the universe is destined calls us toward an interaction, a touching, a give-and-take with the living world. For some, that will mean disconnecting from our screens to reconnect barefoot to the grass, or getting our hands into the soil that feeds us by cultivating a plant or two. To others, it will mean losing awareness of time by contemplating the intricate stem-work of a leaf held against a sunset or working with others in restoring a nearby creek in the watershed one inhabits. Regardless, earthkeeping invites us all to find sacred moments and spaces to breathe slowly and deeply, savoring the air that bonds us with every other living being.

Confessions, Healings, & Forgotten Horizons

To live a life of earthkeeping also demands that we drink from the still waters that can bring healing to our inner violence. To one degree or another, we've all been tainted by subtle forces of darkness; sin has wounded us. As our restless hearts lust wildly after waters that don't satisfy, Babel has set the pace, and we've become complicit "uncreators" in an earthly paradise.

For reasons like these, old and recent Christian traditions have emphasized the importance of confession and self-renewal. At the core of Christianity is a turning-away from evil and a relational encounter with the source of life and healing that permeates and flows through the cosmos—the living God whom Jesus called *Abba*. This encounter is one that takes part in what ancient Eastern theologians named *perichoresis*: the mysterious 'interpenetration' where Jesus is fully 'in God' and God is fully in Jesus. And it is into that divine union—into such sacred ecstasy—that we have all been invited. At the core of Christianity is a universal call to leave behind all darkness and selfishness to become instead adopted family members that receive and extend the extravagant and overflowing presence of God.

Surely the encounter continues to be either downplayed or ridiculed by secularism or truncated by Christianity. Whether Catholic or Protestant, some of today's churches call members to give allegiance to the God revealed in Jesus but are still to embrace practical action toward addressing consumerism and climate change. The split has led people like Wendell Berry to lament how many people of faith continue to be blind or unresponsive to the ecological depravity of today's growth-driven economic system. Will churches exemplify acts of ecological lament and public confession followed by concrete practices that demonstrate true repentance?

One of the ancient oracles of Isaiah spoke of true religion as being one that resulted in working for the good of others. The true worship of Yahweh, said the prophet, must lead to breaking the bonds of injustice and undoing the straps of a yoke, letting the oppressed go free. Then, and only then, will light break forth like the dawn (Isa 58:6).

A New Song, Together

It goes without saying that the gods of Babylon will continue to lure and entice us, reincarnated in ever-new shapes and forms. But the joy that overflows by calling on the name of the Nazarene is the ultimate source of inspiration and resistance. Long ago, Ezekiel, son of Buzi, foresaw a day when the living Yahweh would blow a fresh breath to turn hearts of stone into ones of flesh. Jeremiah spoke of Yahweh writing laws of *shalom* and justice on people's deepest selves. John the seer portrayed the followers of the Lamb as communities of imperfect people being liberated and empowered to live righteously, *together*, inspired by a new song.

Seven chapters ago we met Boris Waterlove, the Doublelives, and the Floating Clowns. Waterlove had an ecological conversion, leading him to realize how caring for the land, the air, and the watersheds is a way of caring for hundreds (millions) of people all at once. With skepticism, and perhaps with some indifference, the Doublelives shrank back. But the Floating Clowns were right on board, feet on the ground. As most insisted on staring down like the overworked citizens of the businessman's worn-out planet, the Clowns bent their heads back to take in the wonder of the stars. They lived differently, inspired by the melody of an eternal tune. They joined the trees that clapped their hands and the mountains that cheered for joy, singing together, well aware of the irony of undertaking a foolish journey that others claimed could not be done.

Sparks in the Dark

We have over 250 years of taking fossil fuels for granted. As of today, around 80 percent of the world's energy comes from polluting sources, and should our economies continue at their current pace, by 2030 we will likely have burnt up to 85 percent of humankind's fossil fuel budget. Today we are living at the expense of our daughters and sons—and of ourselves. We've built an entire global civilization assuming that coal, oil, and gas are infinite resources that magically appeared out of nowhere, and we continue to behave as if our atmosphere can absorb their waste without end.

That is to say that the magnitude and the pace of the change needed in the decade ahead can be disheartening. In fact, it's likely the greatest challenge we *Homo sapiens* have ever faced, making the work of climate justice an uphill journey that's anything but glamorous. And the challenge begs the question: Do we need, then, to take ourselves seriously and enter the climate race with determination?

Creating awareness, organizing communities, implementing technical solutions, standing alongside those who are silenced and suffering, resisting evil and working for good—these are all key stepping stones in the shared path toward climate justice. Taking ourselves seriously and being determined to walk down this narrow road go without saying. Needless to say, it's always challenging to swim against the current, and it's especially challenging to follow after the founder of Christianity because true grace has never been cheap. Heaven's starlight comes, and will always come, at a cost.

But as contradictory as it sounds, the cosmic victory to which the first followers of Jesus testified to is also an easy yoke. For one, we're promised sisters and brothers and all sorts of people of good will with whom to sing and share the burden. We've also been guaranteed that the floodgates of heaven will be opened to us, inspiring and empowering us with the eternal wind that is renewing all creation. We can befriend God. In fact, the Christian experience through two millenia confirms that God struggles with us—that God *suffers* with us—always comforting us with the joyful presence of the Spirit (Rom 8). Along with Abraham Lincoln, Martin Luther King Jr., and Dorothy Day, countless communities of unsung heroines and heroes have found their ultimate source of strength and inspiration in the Most High. "Yahweh is my strength and my shield; my heart trusts in him, and he helps me. My heart leaps for joy, and with my song I praise him" (Ps 28:7).

Surely, the pathway toward climate justice is one riddled with enticements to pursue the glittering voices of beastly powers. The temptations to misplace our desires at the expense of the climate are alive as ever because going against the currents of indifference and comfort will requiring courage, commitment, and endurance instead. But despite the prospects, it remains true that the darkest hour arrives just before dawn, as Costa Rican poet Isaac Felipe Azofeifa once recognized. A narrow window of possibility is still open for us to stretch our eyes into the distant stars and walk in wonder after the Lamb. Even if imperfect and painfully incomplete, our journey will not be in vain. In striding together across dry deserts and dark valleys, we can take courage from the Morning Star that shines in the darkness, because no darkness has ever been strong enough to overcome it. Through winds and storms, the Star will comfort us as we pursue the vast horizons of eternity waiting for us at the end of history. We won't be alone. We'll join and be surrounded by a constellation of witnesses who were foolish enough to say "yes" to the God of life. We'll be comforted by the Living Spirit who can unite us across our differences. We can orient our cities toward sacred relationships of ecological romance, even as our climate of apathy changes into one of faith, hope, and love—despite our fears and limitations, against all odds.

Afterword

I don't know the exact reasons that have made you reach the end of this book, dear reader; but I'd be curious to hear them. You're invited to write to me or to have your say publicly by following the link to the book's website: www.climateofdesire.com.

For now, allow me to end with a confession and a suggestion. While it's likely that the rather novel (and odd) combination of the subjects of this book might have awakened your interest, I hope you and I overcome the irony we've gotten ourselves into. Novelty makes our brains spark, releasing the short-term happiness hormones that are set free when we experience the latest thing. In our culture of entertainment, we're very much addicted to that. Moreover, like writing, reading can make one step back from reality, so to speak, to engage with the world primarily with the mind, at arm's length. It is reasons like these which can turn this book into perhaps the greatest obstacle to advancing toward the horizons it calls us to.

However, it's also true that for every 10,000 words, there's sometimes a deed. So, to go beyond the temptation of feasting on thoughts but lacking in actions, I'd like to suggest a first step. Would you be willing to take sufficient time on an afternoon before sunset to meet with two or three friends and go outside, to a place of quiet where you can see the wide-open skies? No watches. No phones. Just you all.

Once settled there, take a long reflexive look at everything that's around you. Recognize yourselves as living members of a sacred world. Pause to pay close attention to something you find sublime, majestic, radiant. Pause again. Breathe in as if that was your last breath. Breathe out, likewise. And then, when you feel ready, invite the Eternal One to inspire you and indwell you, and, with due pauses, voice out words like these:

A Climate of Desire

Living Source of all there is,
Cosmic Consciousness made human,
Breath of Eternal Life:

How can we best respond to the brokenness of our world and within us?

Heal our inner violence . . .

Disturb our comfort . . .

Empower us to turn away from evil . . .

Enliven our imaginations . . .

Extend the horizons of our hopes . . .

Fill us with wisdom, courage, and love
to work for peace where there is suffering,
to be the change that others claim cannot take place,
to ground ourselves as faithful keepers of this gorgeous planet earth

Annex

An Ongoing Experiment
The Beginnings of a Concrete Response

In 2014, a group of around ten people began to meet informally every few weeks, asking ourselves the question: "What would it look like to live out our faith in our day and age when we face the possibility of climate collapse?"

As of the time of this publication, both personally and collectively, we've had the opportunity to host movie screenings and dialogue sessions, even as we have visited ecologically-threatened sites to intercede and partake in the sufferings of the living world. In turn, some of us have actively participated in town hall consultations, while others have partaken in sustainable food initiatives or become advocates of the renewable energy transition. Not long ago, many hosted a fund-raising concert to support aboriginal peoples on the frontlines. Some have marched in public rallies or participated in beach clean-up activities, and still others have written songs or poems or personal testimonials of what it takes to stand up to beastly powers that are at odds with the integrity of the living world.

Regardless, below you'll find one of the consensual documents containing the identity, mandate, and practices of a citizen group that we eventually called Earthkeepers...

~

Place yourself in a rugged uphill landscape on a foggy day, with the mist making it difficult to know which path to follow ahead. Imagine also a forest, whose life depends on the decaying humus on the topsoil and sufficient rain from above.

This chapter is a map of sorts, outlining some of the basic directions toward a mountaintop from which climate justice flows down like a river. It also lays down some important ingredients that we identified as necessary for the healthy soil biology sustaining an ecological movement that seeks to be grounded in the biblical and Judeo-Christian tradition.

Despite these natural metaphors, this appendix is probably the driest and most abstract part of this book. It may be more of an X-ray scan, revealing the bare bones of a faith-based citizen group that is seeking to work toward broad-level change in regards to climate justice.

Earthkeepers' Story & Identity

In 2014, the elected Canadian federal government allowed foreign oil corporation Kinder Morgan to drill and test out Burnaby Mountain in British Columbia as a potential location for an oil pipeline expansion. A few of us attended the site, joining hundreds of citizens, academics, seniors, and indigenous folk who were all concerned about the repercussions of the project.

The shock of seeing the police forces arresting young and old alike got a few of us together to begin to ask what the role of the church could be in the struggle. After several prayer meetings and several consultations with other faith-based ecological groups, we eventually discerned the need for a distinctive contribution. As it turns out, something new was born to our surprise, and with the birth came the need to clarify the group's identity, its sense of mandate, and a set of practices that would enable us to carry it through. So after some back and forth, and rather unoriginally, we decided to call the initiative Earthkeepers, seeing ourselves as

> ... *an ecumenical group in Vancouver, Coast Salish Territories, living into a biblical vision for ecology, love of neighbor, and climate justice.*

Such identity was (and is) grounded in at least four convictions:

An Ongoing Experiment

I. Being nourished and informed by biblical, theological, scientific, and traditional knowledge

Wanting to avoid the two common extremes of activism void of thought and verbalism void of action, we are convinced that the work of climate justice requires an integrated approach to knowledge. It calls for "faith that expresses itself through love," as the apostle Paul recognized (Gal 5:6). In turn, as opposed to seeing science and religion at odds with each other, we understand them as two dimensions of the same created reality—two very different kinds of maps describing the same landscape. We also acknowledge that the detached modern way of knowing (what philosophers call Cartesian epistemology) has crept into much of Christian thought and practice in the West. Today we have come to think we know something or someone when we grasp it conceptually, with our minds.

To be sure, head knowledge is one dimension of knowing. However, in a biblical sense knowledge and wisdom are manifested through obedience and relationship. To know God and to be wise are qualities that have to do with people's entire lives, reflecting a living communion with the living source of justice and love and with the rest of the earth's living and nonliving creatures. In that sense, traditional or indigenous ways of knowing are something we do well in recovering, even as we leave behind what has often ended up being little more than a rationalized Christianity.

One does not learn to swim by writing or speaking about swimming, but by jumping into the water. Our ecological crisis thus urges us to acknowledge that lip-service does not do justice to the demands of attentive care and tangible protection that the living earth calls for.

II. A multicultural, ecumenical, and post-Eurocentric embodiment of the gospel

Three big words, all related to each other. The first one, multicultural, stems from the radical reality celebrated by the biblical writers where all peoples from all tongues and all nations are called to be united in the Messiah (Gal 3; Rev 7). The Christian good news celebrates the racial, economic, and ethnic diversity in the world. Of all things, this sweeping vision begins with the sobering fact that Jesus himself belonged to the sidelined artisan class of the poor province of Galilee.

As obvious as it may sound, this cross-cultural multiculturalism is at stake when certain monochromatic varieties of (pop culture) Christianity have expanded throughout the globe, thanks to the mind-boggling reach of mass media. However, to be truly ecumenical requires the global church to come to the recognition that all expressions of the faith in the household of God have a seat at the table. The North has contributions to make to the South, even as the South to the North, and the East to the West. Thus a "post-Eurocentric" embodiment of the gospel acknowledges that Jesus and the apostles were grounded in the cosmic story of creation and sociopolitical liberation recorded in the Hebrew Scriptures. It recognizes that it was not in Paris, London, or in Rome, but in Bethlehem where Christ was born.

III. A commitment to operating as a grassroots network

Most churches and other religious institutions nowadays benefit from what is called "charitable status." Donor contributions enter a seemingly all-beneficial cycle: the state is relieved from paying for the social services churches offer and, in exchange, churches provide tax-deductible receipts to their members, who in turn pay lower income taxes.

However, this model often has an unseen side to it. What happens, for instance, when church members or contributing donors are disturbed (or alienated) by the message that comes from the pulpit? Or what to say of churches who fear the loss of their charitable status and thus hold back from being vocal about state failures or abuses of power by government institutions? Or, to sift more thinly, what if the donor money comes from dubious or polluting industries?

Reasons like these lead Old Testament scholar Walter Brueggemann to call the church to reclaiming the freedom of the pulpit. The church needs voices that speak fearlessly and unmask the corporate and political forces that are affronting creation. So to have a greater liberty of speech, we discerned that bearing witness to the living reign of God over all things is better served by operating on a grassroots basis. As far as humanly possible, at Earthkeepers we want to reclaim such freedom by operating as common citizens who work toward climate justice without charge.

IV. A life of personal and communal prayer, worship, and contemplative action

One of the tendencies of liberal Christianity has been to bypass the place of prayer and worship in favor of human action. As sociologist Craig Gay has remarked in *The Way of the (Modern) World: Or, Why It's Tempting to Live As If God Doesn't Exist*, this has even led some branches of the church to attempt to build the kingdom of heaven here on earth by themselves, as if God didn't exist. On the other end, more conservative expressions of Christianity often believe that prayer is either about inner transformation alone (a tradition known as pietism), or about asking God to act in the spiritual realm, as if that did not have any implications for how we live in the physical world (a distortion known as dualism or Gnosticism).

However, the unity and reconciliation of all things calls for the recognition that life is one single continuum with outer and inner dimensions (physical and spiritual, earthy and heavenly, visible and invisible). And prayer and worship are where the veil between these dimensions becomes thin and translucent, where the energies of heaven are breathed into us by the Eternal Spirit, to renew and revitalize us to participate in the healing of the world. Wanting to transform the world without God is puffed up and dangerous; praying and then leaving society unchanged is a comfortable illusion. Instead, the ministry of reconciliation calls us to recognize with Walter Wink that prayer and worship are the ultimate acts of partnership with God.

∼

What has Earthkeepers discerned in light of these four convictions—especially knowing that other faith-based organizations are already engaging with the challenges of climate justice? What unique role might a group like Earthkeepers have?

What Earthkeepers is on about

In recognizing the need to collaborate with other organizations (whether faith-based or not), Earthkeepers has been operating on the conviction of not reinventing the wheel. Instead, the aim is to fill some of the gaps not covered by other groups. Particularly,

We connect, equip and inspire the Christian community to engage with our natural contexts, and cultural and political institutions to tangibly witness to the reconciling love of God for all peoples and all creation.

Convinced of the need to move beyond individual and toward collective action, we seek to be nurtured and inspired by the Spirit to engage the issues at hand *together*. In doing this, we recognize that a cross-shaped political engagement needs to be grounded in a lived experience of the living source who sustains creation. Following the pattern of Jesus and the prophets, we thus realize that faithful action in the *city* needs to be preceded by engaging with the Spirit in the *wilderness*.

But contrary to what has often been rightly labeled as a genetically modified gospel, which narrows the scope of the regeneration to the spiritual and to individual self-realization, the heavenly ecosystem foreseen by the prophets and inaugurated by Jesus is one for all the living world (Isa 40, 55; Rev 1). The cosmic vision of the apostles articulated in the New Testament summons the church to be a community where this reconciliation is practiced and announced to the wider world.

For one, this takes place by embodying the new way of ethnic unity and peace where God is all and in all (Col 3:11; Eph 3:5). For another, it's also a call to speak truth and witness to the kingship of the Lamb before idolatrous peoples and bullying earthly powers (Rev 11; cf. Acts 19:23–41)—be it profit-worshipping corporations, insatiable consumer masses, or short-sighted governments. As such, it's a call to build on the legacy of movements such as the Clapham slave trade abolitionists in the nineteenth century, or of Martin Luther King Jr. and the civil rights advocates in the twentieth.

In particular, the Hebraic dimension of the story of Jesus and the church also summons us to express tangible concern for the local and global poor, who are most drastically affected by the causes and effects of climate change. Likewise, Earthkeepers' mandate is also grounded in the recognition that we human beings are called to be preservers of a world that we share with other creatures.

Living it Out

"Show me how you live, and I'll tell you what you believe in," the saying goes. Which is more or less a variation of the call in the letter of James: "Show me your faith by your deeds" (2:18), or of Jesus' own words "by their fruit you will recognize them" (Matt 7:20).

Beyond speaking about values, at Earthkeepers we have come to recognize the importance of taking on certain practices because values are often abstract and sentimental, but practices are tangible and concrete. Thus we seek to live out our identity and mandate in three main ways.

An Ongoing Experiment

Cultivating Awareness

First, we want to deliver presentations, tell stories, and foster dialogue sessions to connect the dots and create awareness of the moral and theological implications of climate change. Far beyond sharing the facts (which are often quite depressing), we recognize the need to challenge the predominant consumerist myth with the healing story of creation and redemption handed on by the biblical authors. This book seeks to respond to such a need.

Healing through the Arts

A second practice involves facilitating integrative and artistic spaces for healing our relationship to nature and all creatures. At Earthkeepers, we seek to embrace art and storytelling as supreme modes of speech. This has taken on various forms: in climate marches, carrying hand-painted signs with a public prayer, dressed up in sackcloth and ashes for a vigil re-enacting the stations of the cross at the proposed pipeline expansion sites at Burnaby Mountain, using branches and stones in prayer gatherings as symbols of the burdens we lay down. Far beyond seeing art as an object of passive contemplation or selfish self-expression, we are reminded that art uniquely equips us for action as we voice creation's praise. Art is the needle that weaves us back into the fabric of the universe.

Civic & Political Engagement

Thirdly, Earthkeepers has been fostering a sense of compassionate citizenship through prayerful and tangible political engagement. In a day when the voices of entertainment shout as loud as ever—and when going to the mall to entertain one's self and shop until one drops has become the norm—we are grasped by the desire to be consumed by a different vision. In ancient days, being a Roman citizen was a badge of prestige and high social standing. In contrast to such a backdrop the apostle Paul summoned the earliest followers of Jesus to be citizens of heaven here on earth. The call was nothing short of pledging allegiance to a different kingdom, a summons to becoming "living sacrifices" consumed in the work of collaborating with God as the ecosystem of heaven infuses our earthly home with peace, hope, and love (Rom 12; 1 Cor 13; 2 Cor 5; Phil 3).

These three practices are themselves nourished by four underlying streams:

a. Table-fellowship and relational funding. The seeds of what eventually became Earthkeepers were sown during dinner table conversations.

For one, a senior minister confessed to us that pastors are usually "gatekeepers" (and not always "risk-takers") and that most of them have their hands full with administrative work. He then suggested that pastors were not likely the people pushing for change. In turn, we also realized that the fact that churches have charitable status often impedes them from speaking openly on controversial issues (in this case, unmasking the corporate- and state-driven interests that are leading to accelerated levels of global warming). This led us to the conviction that we should find ways of operating that were not financially compromised, so we decided to work without remuneration.

Surely, working this way has come with a cost, as it forces one to move slowly, but that has been a healthy antidote for an age when everything is ready-made. In fact, it has certainly helped to keep things more intimate and relational, especially in a wealthy city that is as suffocated by social isolation as Vancouver is. We've been surprised by finding a sense of community gathered around meals and home meetings as normal citizens who are wanting to embrace a sense of participation in civic affairs.

b. Collaboration with indigenous peoples, especially those who are part of the church family. Heirs of 500 years of the European colonial project, we continue to bemoan the power imbalances that have put indigenous peoples on the underside of history. In truth, at Earthkeepers, we acknowledge it's impossible to recover right relationships with the land without also restoring our relationships with the peoples who have lived in the land and called it home for thousands of years.

We are called to remember that the apostolic ministry of reconciliation implies not only a sense of peace with God, but with other humans and with the creation itself (2 Cor 5; Col 1). It calls for the restitution of what has been taken away, the healing of historical imbalances of power, and a sense of horizontal partnership in a common struggle. It also calls for respecting indigenous efforts at nationhood and self-determination in their own right.

c. Acts of public confession, intercession, and spiritually nurtured direct action in the public sphere. Lecturing on the book of Revelation, biblical scholar N. T. Wright remarked how large chunks of Western Christianity have taken for granted their own position of cultural prestige. Ever since the fourth century when the emperor Constantine made his vows of faith, a tectonic shift in imperial politics led the church away from being persecuted to becoming the persecutor. Once symbols of scorn, crosses became emblems plastered into military banners and shields. Christianity was turned into the religion of victors and conquerors, and to lesser and greater extents, it has remained so ever since. What might it look like to resist bowing one's knee and burning incense in honor of the imperial powers of our day?

A first step is to recognize our complicity in our fossil-fueled system. It calls for acts of personal and corporate confession, followed by intercession and the taking of decisive, creative steps toward a low-carbon economy. (Readers can return to chapter 5 for additional guidance.)

Secondly, spiritual activity in the public realm calls for recovering another dimension often present in marginalized strands of the Christian tradition: that is, the practice of peacemaking direct action and nonviolent civil disobedience. As countless land-protectors and some fellow earth-keepers have demonstrated with their life's testimony, the time is ripe for exploring afresh what this might imply in a day when states and police forces continue to be co-opted by the disguised, totalitarian rule of corporations. (Some of the grounds for this practice were discussed in "Overcoming Evil & Doing Good" in chapter 7.)

d. Seeking to live with much greater ecological integrity. Walking, biking, and taking transit; making homes and buildings less energy-reliant; living on less and liking it more; using community-based tool libraries; participating in the sharing economy; caring for what we have instead of endlessly desiring what we don't—these are all necessary personal lifestyle responses to the climate challenge, whose pros and cons we discussed in chapter 5.

Perhaps nowhere is change more urgent than in what we serve at our tables. Besides the air and water we breathe and drink, food is our most intimate bond with the rest of the living world. In fact, we live through the death of other creatures. Far from fostering a fatalistic attitude, this profound mystery should make us intentional and attentive to what we eat, and who and where we obtain our foods from. Supporting direct-trade and system-change initiatives to strengthen small/medium-scale agroecological farms are key steps toward eating (and living) with greater ecological integrity.

∼

That's more or less it . . . These very broad strokes are at least a sketch of the beginnings of the developing story of Earthkeepers. Being just one among many other spiritual communities who are reexamining the enormous story of Jesus, Moses, and the prophets, Earthkeepers is working to recover that story and reinterpret it in fresh ways, attentive to the local and global realities of our time.

Hopefully, we have somehow inspired you to join in and do likewise as you gather with other like-hearted folks wherever you all may be.

Peace be with you.

Endnotes

Opening Words

1. These are large issues and questions to be explored in detail elsewhere. It's not my intention here to contend intellectually with the secularist worldview. I would simply like to acknowledge, first, that its critique of the abuses of religion is often accurate and true, and secondly, that secularism can often be equally reductionist and tied to a set of modern, materialistic assumptions. While not "religious" as in "theistic," secular ideologies such as Marxism and market capitalism, for example, are both oriented toward invisible ideals they can't ever fully reach and dogmas they can't ever fully explain.

For an accessible but critical engagement with the secularist/humanist tradition, see Alister McGrath, *Intellectuals Don't Need God and Other Modern Myths* (Grand Rapids, MI: Zondervan, 1993). For a recent philosophical engagement with scientism, constructivism, and materialism, see Christian J. Barrigar, *Freedom All the Way Up: God and the Meaning of Life in a Scientific Age* (Victoria, BC: Friesen, 2017). On the history of the development of religion, see Robert N. Bellah, *Religion in Human Evolution: From the Paleolithic to the Axial Age* (Harvard University Press, 2011).

2. Albert Einstein, *Out of My Later Years* (New York, NY: Philosophical Library, 1950), 26.

3. Union of Concerned Scientists, "World Scientists' Warning to Humanity," 1992. On the need for moral and religious traditions to respond to our economic/ecological condition, see Bob Massie, "The Moral Dilemma of Growth," *Reflections*, 2012, available at: http://reflections.yale.edu/article/who-are-we-american-values-revisited/moral-dilemma-growth (accessed May 20, 2017).

4. In Wendell Berry, *The Art of the Commonplace: The Agrarian Essays of Wendell Berry*, ed. Norman Wirzba (Berkeley, CA: Counterpoint, 2002), 306.

5. Classically, Lynn White Jr., "The Historical Roots of our Ecological Crisis," *Science*, 1967, 1203–1207. While White's portrayal may reflect the way some Christians across history have distorted the Bible, his own critical interpretation of what the Scriptures

intend to say is problematic.

Notably, White's portrayal is exaggerated at times and lacks a contextualization of the Genesis creation account in its literary, canonical, and Ancient Near Eastern settings. For instance, White blamed Christianity for making believe that "no item in the physical creation had any purpose save to serve man's purposes" (para. 19). But in making such a remark, he remained silent about the biblical command to serve and preserve the garden in Genesis 2, or about the speeches at the end of the Book of Job that speak of the living world and its creatures as having a value for God quite apart from the value (or lack thereof) they have for humans.

White also blamed Christianity for doing away with pagan animism, thus opening the way for what he saw as the exploitation of "nature in a mood of indifference to the feelings of natural objects" (para. 21). While in some cases that remains painfully true, in chapter 3 we'll glimpse how we humans have *always* been a destructive ecological force, wiping out countless species of megafauna wherever we have settled—whether in Australia as early as 45,000 BCE, or in the Americas beginning around 14,000 years ago.

Chapter 1: Sex & the Cities

6. G. K. Chesterton, *Orthodoxy: The Romance of Faith*. Reprint of 1908 ed. (Chicago, IL: Moody, 2009), 32.

7. The central place of the land in Israelite tradition, and the call to be in right relationship with it, has been carefully explored by Ellen F. Davis in her book, *Scripture, Culture and Agriculture: An Agrarian Reading of the Bible* (Cambridge: Cambridge University Press, 2009).

8. cf. Lev 18:24–25, 28; Num 34:33–34.

9. Walter Brueggemann, *The Land: Place as Gift, Promise, and Challenge in Biblical Faith*, second ed. [Overtures to Biblical Theology] (Minneapolis: Fortress, 2002), 112–13. Brueggemann sees the tensions which surface in the Book of Jeremiah as revolving around three related issues: covenant fidelity to Yahweh, collective remembrance of Israel's history, and land management. Israel's historical amnesia about having been a landless people that were eventually entrusted with the gift and the task of land tenure aggravates their own destiny; see especially 101–22. Brueggemann points out how the land is a gift coming from Yahweh's radical grace. As such, the land is blessed with abundance and fertility. However, the land by itself becomes a source of temptation outside its covenantal nature whereby humans, land, and Yahweh are in constant conversation. In fact, it allows the Israelites to forget the land-giver and take their blessedness for granted. Among other purposes, the Torah was thus given to awaken Israel to its own historical memory, as well as to keep her from coveting, from frantic accumulation, and from the worship of other gods. These boundaries come into play through the observance of the land Sabbaths and care for the landless; see, ibid., 43–65.

10. The populist, premillennial, end-times theology of the Left Behind series continues to be a laughingstock among scholars. More serious versions of premillennialism and other interpretative approaches are evaluated by Craig Koester, *Revelation: A New Translation with Introduction and Commentary* (New Haven: Yale University Press, 2014), 59–65.

11. Richard Bauckham, *The Climax of Prophecy: Studies on the Book of Revelation* (T&T Clark, 2000).

12. David Orr, *Down to the Wire: Confronting Climate Collapse* (Oxford: Oxford University Press, 2009), 7.

13. G. K. Beale interprets "the earth" in Revelation 11:18 as referring to the *inhabitants* of the earth (more specifically, God's people), and not primarily referring to the earth itself; see *The Book of Revelation: A Commentary on the Greek Text* [The New International Greek New Testament Commentary] (Carlisle, UK: The Paternoster Press, 1999), 616. Nevertheless, one may safely extrapolate and affirm that destroying the earthly habitat upon which people depend counts for the same (as our contemporary ecological crisis has proven to be the case).

14. Some of these have been explored by Orr, *Down to the Wire*, especially ch. 6; Michael Northcott, *A Moral Climate: The Ethics of Global Warming* (Maryknoll, NY: Orbis, 2007).

15. Elizabeth Kolbert, "Why Facts Don't Change Our Minds," *The New Yorker*, Feb 27, 2017.

16. Berry, *The Art of the Commonplace*, 318–19.

Chapter 2: "Just Gimme the Facts"

17. On the explicit relationship between fire and purification, see most notably 1 Corinthians 3:10–15 and 1 Peter 1:7. On the groaning of creation and its hope for liberation, see most explicitly Isaiah 55 and Romans 8:18–23.

18. I became aware of such a letter through a summons written by Loren Wilkinson in response to the incidents around Burnaby Mountain; see Loren Wilkinson, "Kinder Morgan Raises Serious Questions for Christians," *Church for Vancouver*, November 27, 2014, available online at: http://churchforvancouver.ca/kinder-morgan-protests-raise-serious-questions-for-christians (accessed Dec 5, 2014).

19. To have a likely chance to keep global warming below a 2°C increase, the International Panel on Climate Change estimates that only an additional 565 $GtCO_2$ can still be dumped into the atmosphere. (Currently the global village is burning an estimated 31Gt per year, giving us between 15–25 years before we surpass that budget.) In turn, the Province of Albeta's oil sands alone account for an estimated 17 percent of the global carbon budget. That budget means two-thirds of the nation's proven reserves, and 83 percent of proven-plus-probable reserves, need to remain underground. See http://350.org/what-does-carbon-bubble-mean-canada-and-tar-sands (accessed Jan 21, 2017) and http://www.theglobeandmail.com/report-on-business/economy/economy-lab/global-carbon-budget-is-a-harsh-reality-check-for-canadian-investors/article15158549 (accessed Jan 14, 2014).

20. "Beyond Petroleum," *The Economist*, Jan 29, 2015; available at: http://www.economist.com/news/americas/21641288-growth-shifting-oil-producing-west-back-traditional-economic-heartland (accessed Aug 20, 2016).

21. Calculation based on the figure of direct and indirect jobs associated with oil sands development, found in the website of the Canadian Association of Petroleum Producers, available at: https://www.canadasoilsands.ca/en/explore-topics/economic-contribution (accessed Apr 20, 2018).

22. Einstein, *Out of My Later Years*, 26. Einstein himself seems to have a conception of God following that of Dutch philosopher Baruch Spinoza, who believed in an anthropomorphic deity who created the laws of the universe and then set it in motion, never

touching it again. In this materialistic conception, the laws of physics are the "thoughts of God."

23. Less and less Christians believe that the creation account in Genesis 1 did in fact take place in six days (or, for that matter, in seven days). As a vast majority of Old Testament scholarship has shown, the primeval history in Genesis 1–11 is the language of poetry and of sociopolitical subversion of ancient mythologies. The seven days are a rhetorical device not a statement of modern scientific fact. The narrator of Genesis is taking the seven-day week that was common to the Israelites in order to infuse it with divine meaning, as a way to speak in a language that would make sense to ancient peoples. Although the "literal" sense reads as six actual days, the *literature* is not concerned with exact spans of time, but with the declaration (among others) that it is the God of creation, and not the gods of Egypt or Babylon, who is the divine author of a good universe. For an accessible but thoughtful engagement with the Genesis creation account, see John Lennox, *Seven Days That Divide the World: The Beginning According to Genesis and Science* (Grand Rapids: Zondervan, 2011), and John Walton, *The Lost World of Genesis One: Ancient Cosmology and the Origins Debate* (Downers Grove, IL: IVP Academic, 2009).

24. I adapt this illustration from Denis Alexander, *Creation or Evolution: Do We Have to Choose?* (Oxford: Monarch Books, 2008).

25. See Alister McGrath, *Christianity's Dangerous Idea: The Protestant Revolution—A History from the Sixteenth Century to the Twenty-First* (New York: HarperOne, 2008), 372–86. McGrath contends that it was Protestantism's emphasis on a literal reading of the book of words (the Bible) which then spilled into the mode of reading the book of works (nature). It was such an intellectual approach which then led to the (modern) empiricist approach in the field of science. Through this shift, natural reality ceased to be seen as a symbol of anything beyond it; instead, nature became seen as natural, desacralized, unaffected by sacred beings or forces. Furthermore, McGrath argues that Protestantism created a new motivation for the scientific study of nature. For people like John Calvin, the beauty and wisdom of God could be experienced and appreciated through that which God had created. Despite the historical irony if seen in retrospect, it was Calvin and his successors who stimulated scientific research by giving it a religious dimension.

26. Those wanting to go somewhat beyond my summary below may take a further look at David Wallace-Wells' recent article "The Uninhabitable Earth," available at *The New York Magazine*'s website.

27. On the connection between climate stability, food production, and rise of civilizations, see Jared Diamond, *Guns, Germs, and Steel: The Fates of Human Societies* (New York: W. W. Norton & Company, 1999), 35–52, 83–92.

28. Unless otherwise specified, the oceanic and atmospheric findings specified here and below are taken from the Intergovernmental Panel on Climate Change, *Climate Change 2014: Synthesis Report. Contribution of Working Groups I, II and III to the Fifth Assessment Report of the Intergovernmental Panel on Climate Change* [Core Writing Team, R. K. Pachauri and L. A. Meyer (eds.)], IPCC, Geneva, Switzerland, 2014.

29. See ibid., 48–52. The fact that there is a *correlation* between rising temperatures and current rising levels of greenhouse gases does not automatically imply a direct *causation*. The earth's past does reveal periods of time when atmospheric CO_2 remained unchanged, while air temperature dropped, as well as times when the air's CO_2 content dropped, while air temperature remained unchanged or actually rose. In the case of the past century, however, the IPCC's climate model comparison between predictions and observations do show an extremely likely relationship between correlation and causation.

Endnotes

Comparing land and ocean scenarios for "warming forcings" that are both naturally only and naturally and humanly induced, the observed rising temperatures in different areas of the world did correlate much better with the scenarios contemplating both human-induced and naturally induced radiation forcings. The warming temperature trend did follow the prediction accounting for human-induced emissions, instead of the predictions accounting for nonhuman factors alone.

30. In framing the situation this way, I am aware of the Climategate criticism of the "hockey-stick" phenomenon originally popularized by climatologist Michael Mann in the late 1990s. This counter-claim affirms that Mann's findings have been picked up (almost religiously) by the scientific and most of the climate community ever since. The allegation is that the link between observed rise in temperatures and measurable GHG rise in the past 30 years have an undeniable correlation.

While there have certainly been flaws pointed out in Mann's initial methodology (e.g., his reconstruction being selective of temperature records that looked like hockey sticks and that were mostly sourced from one small geographic location), I find it unlikely that, after almost two decades of tough push-back and skepticism around Mann's publication, the international scientific community would nevertheless remain committed to controversial claims that would have otherwise been dismissed as false. While there will always be dissenters of every type and shape no matter the issue at hand, the approach of IPCC's latest Working Group I is simply too robust and multi-layered to be easily brushed away as a mere cherry-picking group-think. (On the composition of the working group, see http://www.ipcc.ch/report/ar5/wg1/docs/WG1AR5_FactSheet.pdf [accessed May 12, 2017]).

31. See https://epi.envirocenter.yale.edu/2018-epi-report/climate-and-energy (accessed May 10, 2018).

32. The causes cannot be solely attributed to global warming, as the country itself has allowed poor logging and forestry practices; see "Indonesian Forest Fires on Track to Emit More CO2 than UK," *The Guardian*, October 7, 2015, online at: https://www.theguardian.com/environment/2015/oct/07/indonesian-forest-fires-on-track-to-emit-more-co2-than-uk (accessed Oct 20, 2015).

33. At a rough estimate, more than 200 million people worldwide live along coastlines less than 5 meters above sea level. By the end of the twenty-first century this figure is estimated to increase to 400 to 500 million; see World Ocean Review, "Living with the oceans," available online at: http://worldoceanreview.com/en/wor-1/coasts/living-in-coastal-areas (accessed Apr 05, 2017).

34. International Organization for Migration, "Migration, Climate Change, and the Environment," *IOM Policy Brief*, Geneva, 2009, 1.

35. For an introductory explanation of ocean chemistry and global warming, see "Living with the Oceans – A Report on the State of the World's Oceans" from the World Ocean Review, available online at: http://worldoceanreview.com/en/wor-1/ocean-chemistry/acidification (accessed Apr 10, 2017). On the effect of ocean changes, see IPCC, *Climate Change 2014*, 49–52.

36. These impacts are taken from Norman Wirzba's summary of the ecological prospects related to the global food system in: *Food and Faith: A Theology of Eating* (Cambridge: Cambridge University Press, 2011), 81–86.

37. This trend is calculated with 14,152 populations of 3,706 species monitored across the globe between 1970 and 2012; see WWF, "2016 Living Planet Report—Summary," Gland, Switzerland, 2016, available online at: http://awsassets.panda.org/downloads/lpr_living_planet_report_2016_summary.pdf (accessed May 03, 2017).

38. Gerardo Ceballos, Paul R. Ehrlich, and Rodolfo Dirzo, "Biological Annihilation Via the Ongoing Mass Extinction Signaled by Vertebrate Population Losses and Declines," Proceedings of the National Academy of Sciences of the United States of America, 2017, 114(30).

39. Although not enough if isolated from other factors, income level (and the material benefit it allows for) was found to be the main determinant of happiness; see John Helliwell, Richard Layard, and Jeffrey Sachs (eds.), "World Happiness Report 2015," New York, Columbia University Press, 2015.

40. See William E. Rees and Jennie Moore, "Ecological Footprints and Urbanization," in *Living Within a Fair Share Ecological Footprint*, eds. Robert and Brenda Vale (Oxon, UK: Routledge, 2013), 16–21. Energy land is the land required by "carbon sinks" (such as forests) to absorb the CO_2 resulting from the burning of fossil fuels.

41. New Climate Institute, "Opportunity 2030: Benefits of Climate Action in Cities," Germany, 2018, 16.

42. Rees and Moore, "Ecological Footprints and Urbanization," 7.

43. Jennie Moore, "Ecological Footprints and Lifestyle Archetypes: Exploring Dimensions of Consumption and Transformation Needed to Achieve Urban Sustainability," *Sustainability*, 2015, 7, 4747–4763. See also: Environmental Protection Agency's Greenhouse Gas Equivalencies Calculator. Available online at: https://www.epa.gov/energy/greenhouse-gas-equivalencies-calculator (accessed Feb 20, 2018).

44. Rees and Moore, "Ecological Footprints and Urbanization," 4.

45. Ronald Wright, *A Short History of Progress* [CBC Massey Lectures] (Toronto: House of Anansi Press, 2004).

46. Rees and Moore, "Ecological Footprints and Urbanization," 7.

47. Yvon Chouinard and Vincent Stanley, *The Responsible Company: What We've Learned from Patagonia's First 40 Years* (Ventura, CA: Patagonia, 2012), 14.

48. These are a few of the many ecological case studies recorded in Diamond's panoramic account, *Collapse: How Societies Choose to Fail or Succeed*, (New York: Penguin, 2005), especially ch. 14.

Chapter 3: Babel, Babylonia, & California

49. See, also, "The Golden Age that Never Was" in Jared Diamond, *The Rise and Fall of the Third Chimpanzee: How Our Animal Heritage Affects the Way We Live*, 285–303 (London: Vintage, 2002).

50. Diamond, *Guns, Germs, and Steel*, ch. 1.

51. Cf. Diamond, *The Rise and Fall of the Third Chimpanzee*, 285–303.

52. Gary Gardner, "Cities in the Arc of Human History: A Materials Perspective" in *Can a City Be Sustainable?* [State of the World 2017] (Washington, DC: Island, 2017), 11–25.

53. Wright, *A Short History of Progress*, 113–15.

54. Niall Ferguson, *Civilization: The West and the Rest* (New York: Penguin, 2011), 5, 205. Ferguson is well aware that global warming is a looming threat (and one of the unfortunate consequences of the West's supremacy). He also recognizes that without the existence of the consumerist society, the Industrial Revolution would have failed considerably. In the absence of competitive markets and freedom of rights to dress and eat and drink as one pleases, there would not have been enough purchasing demand to

meet the increased output of the Revolution's steam-powered machines. Ferguson rightly admits that cheap, available coal was one of the two key factors that enabled Britain to industrialize and gain power over the rest of the world. Still, Ferguson's 347 dense pages leave little room for stressing the centrality of cheap energy in the making of the modern world. Words such as "coal" appear only 10 times in the entire book, "oil" (7), "petroleum" (1), "natural gas" (1), "petrol" (0), "fossil fuels" (0), and "fuel" (0). Like many in modern society, Ferguson is certainly well trained in his own discipline; however, his omissions give the impression that he may not be as skilled in other fields, such as urban ecology, ecological economics, or the fundamentals of thermodynamics.

55. Ibid., 200.

56. In *Afterburn: Society Beyond Fossil Fuels* (Gabriola Island, BC: New Society, 2015), 33, Richard Heinberg underlines this by drawing on Marvin Harris's *Cultural Materialism: The Struggle for a Science of Culture*, an analysis of how the fates of societies are intrinsically bound up to their "ecological infrastructure." Ecological infrastructures (such as fossil fuels) are the underlying conditions that ultimately permit or forbid the growth of human techno-systems. They are the controlling factor that allows or limits societal growth. In the case of agrarian societies, the ecological infrastructure consisted of firewood, grass, field crops, some water power, wind for sails, and animal power for transportation. In the case of industrial societies, these renewable inputs have been largely replaced by exponentially depleting a limited amount of polluting and nonrenewable coal, oil, and gas.

57. Cf. Erik Brynjolfsson and Andrew McAfee, *Le deuxième âge de la machine : Travail et prospérité à l'heure de la révolution technologique*, trad. C. Jaquet (Paris: Odile Jacob, 2015), 9–17.

58. Matthew Stilwell, "Climate Debt—A Premier," *Development Dialogue* 61, September 2012, 42; cf. Global Carbon Project emissions data, available online at: http://cdiac.ornl.gov (accessed Dec 27, 2017). The major emitters since the Industrial Revolution can also be found at http://visuals.datadriven.yale.edu/climateaction.

59. Ferguson, *Civilization*, 199.

60. Important religious, economic, and demographic revolutions that went with these shifts are laid down by R. H. Tawney, *Religion and the Rise of Capitalism: A Historical Study* (Middlesex: Penguin, 1926).

61. Cf. "The Late Middle Ages, ca. 1300–1500" in Sarah Williams, *History of Christianity I*, CD, Vancouver, Regent College, 2008.

62. Consider an interesting stat available in Google Book's Ngram Viewer: the percentage of occurrences of the word "individual" begin to take a significant boost come the decade of 1750–1760, the years surrounding the official birth of the Industrial Revolution.

63. Charles Taylor, *Modern Social Imaginaries* (Durham, NC: Duke University Press, 2004), 49–67.

64. Ibid., 49–67. Taylor acknowledges that figures such as the Hebrew prophets do present a "revisionary stance" toward the human good, calling into question the received structures and myths of society. While early religion (including aboriginal religion) involved an acceptance of the order of things, later religions, including later Judaism and Christianity, set themselves to "repair" the original rift or the original loss. Also, Taylor notes how the urban-centered milieu, in which axial and post-axial religions developed, does open what he calls "new possibilities" of disembedded religion; that is, the search for relating to the divine by revising the notions of flourishing, now understood

as ordinary human well-being. Thus, individuals on their own can relate to the divine in new kinds of "sociality" that are unlinked to the established social order. Likewise, Taylor acknowledges how embryonic state structures contributed to the disembedding, given that they required controlling religious and social life to serve more immediate ends.

65. Cf. Taylor, *Modern Social Imaginaries*, 61–67; Eric Assadourian, "The Rise and Fall of Consumer Cultures" in *Transforming Cultures: From Consumerism to Sustainability* [State of the World 2010] (Washington, DC: Island, 2010).

66. Here and following, I unapologetically mimic and borrow, often literally, from the eloquent storyline of fossil fuels presented by the Post Carbon Institute in "The Ultimate Roller Coaster Ride: An Abbreviated History of Fossil Fuels," available online at: https://youtu.be/cJ-J91SwP8w (accessed Oct 20, 2016).

67. The way governments and corporations have used psychoanalysis as a powerful means of persuasion is narrated in Adam Curtis's TV miniseries entitled *Century of the Self*.

68. Alongside Coca-Cola, Ferguson sees jeans as the archetypal evidence of Western superiority; see, *Civilization*, 196–255.

69. Eric Assadourian, "The Rise and Fall of Consumer Cultures," 14.

70. Eric Assadourian, "Re-Engineering Cultures to Create a Sustainable Civilization," in *Is Sustainability Still Possible?* [State of the World 2013] (Washington, DC: Island, 2013), 117.

71. The religious dimensions of consumerism have been thoughtfully explored by William Cavanaugh, *Being Consumed: Economics and Christian Desire* (Grand Rapids: Eerdmans, 2008) and James K. Smith, *Desiring the Kingdom: Worship, Worldview, and Cultural Formation* [Cultural Liturgies, vol. 1] (Grand Rapids: Baker Academic, 2009).

72. Assadourian, "The Rise and Fall of Consumer Cultures," 3–16; cf. Kate, "A New Hedonism: A Post-Consumerism Vision," *The Next System Project*, 22 November 2017, available at: https://thenextsystem.org/learn/stories/new-hedonism-post-consumerism-vision (accessed March 20, 2018).

73. Norman Clarck, "Home Alone with Technology: An Interview with Neil Postman," available online at: https://ir.uiowa.edu/cgi/viewcontent.cgi?article=1402&context=ijcs (accessed June 30, 2018).

74. Niall Ferguson, *Empire: How Britain Made the Modern World*, reprint ed. (New York: Penguin, 2004).

75. A term used by Walter Wink, *Unmasking the Powers: The Invisible Forces that Determine Human Existence* (Philadelphia: Fortress, 1986), 50.

76. E. F. Schumacher, *Small is Beautiful: Economics as if People Mattered*, reprint of 1973 ed. (New York: Harper Perennial, 2010), 32.

77. Madan Sarup, *An Introductory Guide to Post-Structuralism and Postmodernism*, 147. According to Sarup, this perpetual present leads to hedonism, a lack of social identification and obedience, and selfishness.

78. Gerhard von Rad, *Genesis: A Commentary*, revised edition [The Old Testament Library] (Philadelphia: Westminster John Knox Press, 1973), 63.

79. Cf. Iain Provan, *Discovering Genesis: Content, Interpretation, Reception* [Discovering Biblical Texts] (Grand Rapids: Eerdmans, 2015).

80. In the shape and feel of the following remix, I'm indebted to the sort of hermeneutical approach endorsed by Brian Walsh and Sylvia Keesmaat in *Colossians Remixed: Subverting the Empire* (Downers Grove, IL: Intervarsity, 2004), and Walter Brueggemann in *Cadences of Home: Preaching among Exiles* (Louisville: John Knox, 1997). In regards

to the historical grounds for some of my claims, I borrow from John H. Walton's *Ancient Near Easter Thought and the Old Testament: Introducing the Conceptual World of the Hebrew Bible* (Grand Rapids: Baker Academic, 2006), and Wes Howard-Brook's, *"Come Out, My People!": God's Call Out of Empire in the Bible and Beyond* (Maryknoll, NY: Orbis, 2010), 42–51.

Chapter 4: Climate Change & the Good News

81. See Jer 30–31; cf. Ezek 36–37; Joel 2; et al.

82. On the relationship between the ministry of Jesus and the renewal of Israel understood as peasant-critique movement, see Richard Horsley, *Jesus in Context: People Power, and Performance* (Minneapolis: Fortress, 2008), especially 35–55; N. T. Wright, *Jesus and the Victory of God* [Christian Origins and the Question of God, vol. 2] (Minneapolis: Fortress, 1997), especially 320–82, 477–539.

83. Isa 43; Rom 1:1–5; 15; Col 1:15–23. 2 Cor 5:17.

84. Berry, *The Art of the Commonplace*, 319.

85. Again, the central place of the land in Israelite tradition, and the call to be in right relationship with it, are carefully explored by Davis, *Scripture, Culture and Agriculture*, and Brueggemman, *The Land*.

86. In a survey and assessment of historical approaches to the Old Testament found in his *Old Testament Ethics for the People of God* (Downers Grove, IL: Intervarsity, 2004), 387–412, Christopher Wright highlights a few reasons why the Old Testament's "earthly" and political dimensions have been sidelined since early on in Western history. To begin with, one could recall the second century ship-master and church leader Marcion of Sinope, who devalued the books of the "old" testament in favor of only following selected documents of the "new." (Marcion was eventually considered a heretic, which comes from the Greek word *airesis*, meaning "things chosen.") In turn, fourth-century emperor Constantine turned Christianity into a religion of conquest, creating power-thirsty fans, on the one hand, but also radical skeptics about faith's relationship to politics, on the other. Largely under the influence of scholastic theologians, the Middle Ages in Western Europe eventually turned the faith into a highly dogmatic discipline detached from worldly affairs. Later in the sixteenth century, Martin Luther struggled with the Judaic roots of Christianity, leading him to overemphasize "grace" over "law," even as persecuted Anabaptist Christians were marginalized to the point where they came to place an excessive emphasis on personal morality at the expense of social transformation. In the nineteenth and twentieth centuries, many dispensationalists prided themselves in their ability to map down the history of salvation with mathematical precision, affirming the need to save souls in the age of grace, but downplaying the need to press toward the healing of our world today, often believing it was all going to be burned up.

87. Cf. Craig Gay, *The Way of the (Modern) World; Or, Why It's Tempting to Live as if God Doesn't Exist* (Grand Rapids: Eerdmans, 2008), 181–232.

88. Cf. Lesslie Newbigin, *Foolishness to the Greeks: The Gospel and Western Culture* (Grand Rapids: Eerdmans, 1986), 21–41, 65–94. This is obviously a partial description, because the Enlightenment's twin brother was nineteenth-century Romanticism, which elevated human desire as the key to unlock the mysteries of nature. There were also demographic reasons related to the increasing urbanization and to the breaking of traditional, religious, and agrarian bonds that came with it.

ENDNOTES

89. Cf. Michael J. Perry, *Under God?: Religious Faith and Liberal Democracy* (Cambridge: Cambridge University Press, 2003), vii-xi. Besides having learned our lesson from the Thirty Years' War, Perry remarks how secularity is equally divisive, fundamentalist (in some issues even more so than religion), and can equally be an imposition of beliefs (and assumptions); see 35-52, and also cf. Glenn Tinder, *The Fabric of Hope: An Essay* (Grand Rapids: Eerdmans, 1999), 196-201.

90. Quote from Morris's *News from Nowhere*, cited in Wright, *A Short History of Progress*, 120.

91. This is the vision recorded in Ezekiel 47:1-12 and taken up again in Revelation 21-22.

92. In this section I draw from an unpublished paper by Rikki Watts, "On the Edge of the Millennium: Making Sense of Genesis 1." Watts's comparison of Israelite and other ancient creation myths shows how the writers of Genesis were both affirming and subverting the creation myths of their neighboring nations. Especially noteworthy is the similarity of Israel's creation account with Egyptian mythology, even if Egyptian stories were still theogonic (that is, concerned with the emergence of the gods as personifications of aspects of nature). In contrast, the biblical creation account is interested in presenting Israel's god Yahweh as supreme over the god(s) of Egypt, in presenting all of humanity (not only the kings) as made in God's image, and in "de-personifying" nature and its forces.

93. See Walton, *Ancient Near Eastern Thought and the Old Testament*. In understanding the relationship between social symbols and the use of power (and aware of the risk of anachronisms), I also draw tacitly from the power dynamics examined by S. R. F. Price, *Rituals and Power: The Roman Imperial Cult in Asia Minor*, rev. ed. (Cambridge: Cambridge University Press, 1985).

94. Hebrews 8:1-13. Comparing the ministry of Moses' tabernacle and Solomon's temple with the fulfillment of the new covenant foreseen by Jeremiah, the writer affirms: "They serve as copy and as shadow of the heavenly things.... In speaking of a new covenant, he makes the first one obsolete. And what is becoming obsolete and growing old is ready to vanish away" (8:5, 13).

95. Cf. Francis Watson, *Text and Truth: Redefining Biblical Theology* (Grand Rapids: Eerdmans, 1997), 277-304.

96. Revelation 21-22. Cf. Gordon Fee, *Revelation* [New Covenant Commentary Series] (Eugene, OR: Wipf & Stock, 2011), 289-306. The relationship between the garden, the temple, and the mission of Jesus and the church is thoroughly explored by G. K. Beale in *The Temple and the Church's Mission: A Biblical Theology of the Dwelling Place of God* [New Studies in Biblical Theology] (Downers Grove, IL: IVP Academic, 2004). See also this chapter's final endnote below.

97. The rationale and implications of the Israelite social and ecological paradigm are presented by Wright, *Old Testament Ethics*, 23-102, 103-144, respectively.

98. In "On the Edge of the Millennium," Watts identifies the serpent in Genesis 3 as a symbol of Egypt, associating it with Pharaoh's crown, which carried Urea, an enraged female cobra that functioned both as a symbol and the actual repository of Egypt's power. Regardless of the validity of the association, the symbol should not be pressed too far, as its plasticity gives it the ability to speak into multiple contexts all at once.

99. Walter Brueggemann sees Jeremiah as a pivotal figure in the history of ancient Israel, understanding him as the character who uniquely interweaves the discontinuous and yet continuous transition between the covenant of Moses, Israel's exile in Babylon,

and the future promise of a new covenant; see *The Theology of the Book of Jeremiah* [Old Testament Theology] (Cambridge: Cambridge University Press, 2006).

100. Mark 11:15–19 and parallels.

101. Mark 11:20–25; 12:9–11; 13:1–31; John 1:14. N. T. Wright has argued convincingly that the temple is the key symbol to understanding the person and mission of Jesus and of Jesus' movement; see N. T. Wright, "Jesus and the Identity of God," *Ex Auditu*, 1998, 14, 42–56; and in more detail, N. T. Wright, *Jesus and the Victory of God*, 405–27, 489–509.

102. The earthly feel of the political life and ministry of Jesus (which I have sought to portray here) has been vibrantly captured in dramatic narrative by Gerd Theissen, *The Shadow of the Galilean: The Quest for the Historical Jesus in Narrative Form*, trans. John Bowden (Philadelphia: Fortress, 1986).

103. See John D. Crossan and Jonathan L. Reed, *In Search of Paul: How Jesus' Apostle Opposed Rome's Empire with God's Kingdom* (New York: HarperCollins, 2004), 267–269

104. Some of the political dimensions of the Bible have been explored succinctly by Richard Bauckham, *The Bible in Politics: How to Read the Bible Politically* (Louisville: John Knox, 1989). For the political connotations of the crucifixion, see 142–50.

105. For a thoughtful debate around the events and meaning of Jesus's life, death, and resurrection, see Marcus Borg and N. T. Wright, *The Meaning of Jesus: Two Visions*, 2nd rev. ed. (New York: HarperOne, 2007).

106. This quote is presumably found in Tertullian's *De carne Christi* 5. 4; although his authorship is contested.

107. John 20:19–23; 1 Cor 3:16–17; Eph 2:20–22; 1 Pet 2:4–10; Rev 11; cf. N. T. Wright, *The Challenge of Jesus: Rediscovering Who Jesus Was and Is* (Downers Grove, IL: Intervarsity, 1999), 126–49. In G. K. Beale, "Eden, the Temple, and the Church's Mission in the New Creation," *J. Ev. Theol. Soc.*, 2005, 48(1), the author notes: "Christ, as the Last Adam and true king-priest, perfectly obeyed God and expanded the boundaries of the temple from himself to others (in fulfillment of Gen 1:28). We are to continue that task of sharing God's presence with others until the end of the age, when God will cause the task to be completed and the whole earth will be under the roof of God's temple, which is none other than saying that God's presence will fill the earth in a way it never had before. This cultic task of expanding the presence of God is expressed strikingly in Revelation 11. There the Church is portrayed as a 'sanctuary' (vv. 1–2), as 'two witnesses' (v. 3), and as 'two lampstands' (v. 4), the latter image of which, of course, is an integral feature of the temple. The mission of the Church as God's temple is to shine its lampstand like light of witness into the dark world" (31).

Chapter 5: Giving to Caesar What is God's?

108. So argues Orr in *Down to the Wire*, 84–92, who upholds Lincoln as a helpful example for those seeking to respond to today's climate crisis. It's important to be mindful too that Lincoln was partially pushed to his stance, like most politicians, by a growing public opinion against slavery, especially in the North, and because of the political necessity of the Civil War; see https://libcom.org/history/lincoln-emancipation-howard-zinn (accessed June 03, 2017).

109. Jim Forest, *All is Grace: A Biography of Dorothy Day* (Maryknoll, NY: Orbis, 2011).

ENDNOTES

110. "An Evangelical Movement Takes on Climate Change," *Newsweek*. March 9, 2016. Available online at: http://www.newsweek.com/2016/03/18/creation-care-evangelical-christianity-climate-change-434865.html.

111. For one, Herod Antipas was a client king sponsored by Rome and members of the Sanhedrin themselves were involved in the collection of Roman tribute. Like elsewhere around the Roman Empire, the establishment maintained its supremacy by having local elites to their side and then managing local institutions through them; see Peter Oakes, "Roma" in *Diccionario de Jesús y los Evangelios*, ed. Joel B. Green et al. (Barcelona: Clie, 2016), 1018–20.

112. Cf. Richard Horsley, *Covenant Economics: A Biblical Vision of Justice for All* (Louisville: Westminster John Knox, 2009), 82–83, 130–32.

113. See Wright, *Jesus and the Victory of God*, 502–7. For a contrasting perspective arguing for Jesus' (mild) acceptance of Roman taxation, see Bauckham, *The Bible in Politics*, 73–84. Bauckham argues that Yahweh's claim on his people did not conflict with Caesar's right to receive taxation. He grounds this suggestion in the Old Testament distinction between "the things of God" and "the things of the king" (1 Chr 26:32). To be sure, in biblical thought, God's rule does not immediately exclude Caesar's, and Caesar may have a constrained right. Moreover, some even argue that Jesus' statement might encompass the recognition that the "things of Ceasar" belong to God (cf. Ps 24:1), or that one can be both faithful to God and to Ceasar, as long as one shows loyalty *and* deference to the former. What is clear, however, is that nowhere in the Scriptures does one find allowance for people in power to do as they please. For one, the *Pax romana* would have also been a direct affront to the imminent expectation of the coming kingdom of God, perhaps nowhere as clearly predicted as in the Book of Daniel. For another, the Book of Revelation repeatedly blows a whistle against the arrogance of autonomous powers who forsake justice or oppress others. Such was the case of Tiberius's rule exercised through Pilate, Herod Antipas, and the temple authorities. Reasons like these make an interpretation like Wright's more plausible in lining up with the broader biblical tradition, which recognized that all kings and queens are called to exercise their authority humbly, under God, for the good of others.

114. Compare this subsection with Revelation 13 and Richard Bauckham, *The Theology of the Book of Revelation* [New Testament Theology] (Cambridge: Cambridge University Press, 1993), 54–65.

115. Comparison taken from Heinberg, *Afterburn*, 33.

116. Sustainable Canada Dialogues, "Acting on Climate Change: Solutions from Canadian Scholars," 21.

117. McGill University, "How Much are We Using?," available online at: https://www.mcgill.ca/waterislife/waterathome/how-much-are-we-using (accessed May 13, 2017).

118. Figures estimated for 2012; see Statistics Canada, "Waste Disposal by Source, Province and Territory," available online at: http://www.statcan.gc.ca/tables-tableaux/sum-som/l01/cst01/envir25c-eng.htm (accessed Jul 03, 2016).

119. This data stems from the Carbon Disclosure Project's "CPD Carbon Major's Report 2017," 8. The report (p. 5) states that 100 extant fossil fuel producers can be linked to 923 gigatons of carbon dioxide-equivalent (GtCO2e), which came from direct operational- (producer side) and product-related (consumer side) carbon dioxide and methane emissions (1854–2015). Those 100 companies are responsible for over half (52 percent) of global industrial GHGs since the dawn of the industrial revolution in 1751. In addition, the report states that "upstream," consumer-side emissions (Scope 3)

account for over 90 percent of the burning of fossil fuels "downstream." This could make one believe that the great weight of responsibility falls on the consumers' end. However, best practices in sustainability leadership uphold the principle of "extended-producer-responsibility." That is, even if the responsibility is shared with all of us who consume the energy, the greater burden falls on the companies that made the energy available in the first place.

120. Moreover, the chief activities responsible for Canada's overall increase of GHG emissions since 1990 are "Road Transport" (40 percent increase since 1990) and "Mining and Upstream Oil and Gas Production" (246 percent increase). See Environment and Climate Change Canada, 2014, National Inventory Report 1990–2014: Greenhouse Gas Sources and Sinks in Canada: Executive Summary, available online at: http://www.ec.gc.ca/ges-ghg/ (accessed Jul 03, 2016). The report notes: "In 2014, emissions from Mining and Upstream Oil and Gas Production were more than twice their 1990 values. This trend is consistent with a 91 percent increase in total production of crude oil and natural gas over the period, largely for export, which has grown by over 200 percent."

121. Cf. Michael Maniates, "Individualization: Plant a Tree, Buy a Bike, Save the World?," *Global Environmental Politics*, 2001, 1(3), 31–52.

122. Wayne Visser, *The Age of Responsibility: CSR 2.0 and the New DNA of Business* (West Sussex: John Wiley & Sons, 2011), 220.

123. To make these efforts locally relevant, churches and faith-based groups could create a collaborative, "marathon-to-the-top" game where different churches in the same town or city can compare each other as they work toward a "green crown" of sorts.

124. For initial paths of action on building green teams, see https://www.greenbiz.com/news/2009/05/05/how-build-green-team-first-step-sustainability (accessed Apr 28, 2017) or http://c.ymcdn.com/sites/www.gmicglobal.org/resource/resmgr/docs/gmic_green_team_02.pdf (accessed Apr 28, 2017).

125. This is not to say that eco-champions should endorse the logic of cost/benefit that is so prevalent in today's organizational decision-making, because it's precisely that sort of logic which needs to be challenged. That said, sustainability can and often does have a set of economic benefits, some of which are outlined in Canadian Business for Social Responsibility's Transformational Company Guide. The guide also outlines paths of action and promising case studies of companies taking sustainability seriously; see http://cbsr.ca/transformationalcompany (accessed Apr 29, 2017).

126. Annie Leonard, "Moving from Individual Change to Societal Change" in *Is Sustainability Still Possible?*, 245

127. World Economic Forum, "What is the Cost of Delaying Climate Action?," February 2015, available online at: https://www.weforum.org/agenda/2015/02/what-is-the-cost-of-delaying-climate-action (accessed Mar 03, 2016).

128. Explored in more detail in Klein, *This Changes Everything*, 120–60, 230–56.

129. Premised on a leader's ability to argue and to convince, the call for bold leadership within structures of power is upheld by Orr, *Down to the Wire*, 84–108.

130. I've borrowed and adapted these three recommendations from Citizens for Public Justice, "A Public Justice Vision for Canada's Action Plan," Ottawa, June 2016, available online at: https://cpj.ca/sites/default/files/docs/files/CPJVisionforClimateActionPlan.pdf (accessed Aug 20, 2016). More specifically, these measures call for a reduction of GHGs through price on carbon and regulation of carbon intensive sectors. This price has to be high enough to actually reduce emissions. The recommendation is at least \$30/ton CO_2eq, to increase to at least \$160/ton CO_2eq by 2030. The new carbon economy should

be marked by low-carbon technologies and renewables, energy efficiency at all levels, a massive shift toward compact communities, and public transportation. Likewise, providing justice for the disenfranchised implies funding domestic adaptation toward climate change-related impacts, creating social supports and retraining for people employed in carbon-intensive industry, and funneling funds to the UN Green Climate Fund in order to increase the climate financing by $4 billion per year. Here I'm indebted to Anna-Liisa Aunio from Dawson College in Montreal for pointing out the importance of avoiding the creation of policies that might create a backlash, having affected the less-advantaged disproportionately. Heating homes or transit, for example, are activities whose demand for fossil fuels is fairly inelastic, so simply introducing a carbon tax would not likely reduce GHG emissions significantly. On the one hand, this calls for avoiding a flat carbon tax that disproportionately affects the poor; on the other, it calls for subsidizing renewals from the get-go, instead of reimbursing people after the fact. It's necessary to implement a handy alternative simultaneously.

131. Thomas Princen et al., "Keep Them in the Ground: Ending the Fossil Fuel Era," in *Is Sustainability Still Possible?*, 162–63.

132. For these and other insights in this final chapter section, I'm indebted to University of McGill PhD student Jennifer Gobby and to Assistant Professor Hamish van der Ven.

133. David Korten, *When Corporations Rule the World*, 20th Aniv. Ed. (Oakland: Berret-Koehler, 2015); Joel Bakan, *The Corporation: The Pathological Pursuit of Profit and Power* (New York: Free Press, 2004).

134. Cf. Dror Etzion et al., "Unleashing Sustainability Transformations through Robust Action," *Journal of Cleaner Production* xxx (2015), 1–12. Arguing similarly, on religious grounds, see Franciscus, Enciclical Letter *Laudato Si: On the Care of Our Common Home* (Rome: Vatican, 2015), 166, 169.

135. Andrew Cumbers, "Remunicipalization, the Low-Carbon Transition, and Energy Democracy," in *Can a City Be Sustainable?* [State of the World 2016] (Washington: Island, 2016), 275–90.

136. Cf. Ronald Heifetz, *Leadership Without Easy Answers* (Cambridge, MA: Harvard University Press, 1998), 67–180; Melissa Leach, "Pathways to Sustainability: Building Political Strategies," in *Is Sustainability Still Possible?*, 234–43.

137. Cf. Tom Prugh and Michael Renner, "Cities and Greenhouse Gas Emissions: The Scope of the Challenge," in *Can a City Be Sustainable?*, 77–89; Hamish van der Ven, "Power and Authority in Global Climate Governance," *Global Environmental Politics*, 2016, 16(4) 130–35.

138. To an extent, these three suggested routes of action mirror the priorities suggested by Paul Hawken in *Drawdawn: The Most Comprehensive Plan Ever Proposed to Reverse Global Warming* (New York: Penguin, 2017). Hawken's research team identifies the following as the top ten priorities in terms of their potential of reversing GHG concentrations: 1) refrigerant management, 2) onshore wind turbines, 3) reduced food waste, 4) plant-rich diet, 5) tropical forests, 6) educating girls, 7) family planning, 8) solar farms, 9) silvopasture, 10) rooftop solar.

139. Desmond Tutu, "We Need an Apartheid-style Boycott to Save the Planet," The Guardian, April 10, 2014, available at: https://www.theguardian.com/commentisfree/2014/apr/10/divest-fossil-fuels-climate-change-keystone-xl (accessed Dec 20, 2016).

140. For a fully updated list and numerous case studies on divestment, see https://

gofossilfree.org/commitments (accessed Jul 18, 2018).

141. Direct food sales in the US jumped 55 percent between 2002 and 2007 alone, with over 6,130 farmers markets already operating by mid 2010. New urban food trends have been explored by Peter Ladner, *The Urban Food Revolution: Changing the Way We Feed Cities* (Gabriola Island, BC: New Society, 2011). A deeper and more promising discussion on the alternative food systems that we need to aim for is presented in *Food Sovereignty: Reconnecting Food, Nature & Community*, Hannah Wittman, et al., eds (Halifax, NS: Fernwood, 2010).

Chapter 6: When the Climate Changed

142. World Business Council for Sustainable Development, "Vision 2050: Una nueva agenda para las empresas – Resumen Ejecutivo," Ginebra, 2010.

143. Programa Estado de la Nación, "Desafíos del Desarrollo Humano Sostenible" en *Cuarto Informe de la Región en Desarrollo Humano Sostenible* (San José, Costa Rica: Programa Estado de la Nación, 2011), 363–64. Surely the reasons for these migrations cannot be wholly attributed to climate change, as other interrelated factors such as the land grabs by agribusiness and neoliberal structural economic policies also come into play. These risks are confirmed elsewhere; e.g., World Economic Forum, *Global Risks 2013 - Eight Edition*, ed. Lee Howel (Geneva, 2013). In the case of Latin America as a whole, the highest impact risks most likely to occur during the next decade are: severe food and income disparity, failure to cope with climate change, water shortages, and the effects of rising greenhouse gas emissions.

144. A phrase used by Pope Francis in his latest encyclical, "*Laudato Si.*"

145. Quoted by Richard Bauckham, who thoroughly expounds these historical developments in "Modern Domination of Nature: Historical Origins and Biblical Critique," in R. J. Berry, ed. *Environmental Stewardship: Critical Perspectives—Past and Present* (London: T&T Clark, 2006), 32–50.

146. Cf. Lesslie Newbigin, *The Gospel in a Pluralist Society* (Grand Rapids: Eerdmans, 1989), 220.

147. This and similar stats are summarized by William Rees, "Avoiding Collapse: An Agenda for Sustainable Degrowth and Relocalizing the Economy," *Climate Justice Project*, Canadian Centre for Policy Alternatives, June 2014, 1–3.

148. Estimation by Princeton ecologist Stephen Pacala, quoted by Eric Assadourian, "The Rise and Fall of Consumer Cultures," 6.

149. Ibid., 14

150. Zygmunt Bauman, *Postmodernity and Its Discontents* (New York: New York University Press, 1997), 81–94, 199–208.

151. In this case, Christian ethicist Tim Gorringe, "The Principalities and Powers," in Peter Heslam, ed., *Globalization and the Good* (Cambridge, UK: SPCK, 2004), 79–94.

152. See Loren Wilkinson, "Kinder Morgan Protests Raises Serious Questions for Christians."

153. Wink, *Unmasking the Powers: The Invisible Forces that Determine Human Existence* (Philadelphia: Fortress, 1986), 50. Here one ought to be cautious of finding demons under every unturned stone. Instead, one must always be mindful that some of the evil in the world does proceed from the human heart, calling us to be healed from our inner darkness.

Endnotes

154. Walter Wink, *Engaging the Powers: Discernment and Resistance in a World of Domination* (Minneapolis: Fortress, 1992), 9.

155. A vocation motivated by service to and love for our neighbors, while undertaken in prayer for the public authorities and mockery of the false gods that claim a stake on our cultures; so, Wright, *Old Testament Ethics for the People of God*, 241–43.

156. Vaclav Havel, "The Need for Transcendence in a Post-Modern World." I've replaced Havel's masculine nouns and pronouns to make the statement gender-inclusive.

157. Christopher J. H. Wright, *The Mission of God: Unlocking the Bible's Grand Narrative* (Downers Grove, IL: IVP Academic, 2006), 178.

158. Cf. Rodney Stark, *For the Glory of God: How Monotheism Led to Reformations, Science, Witch-Hunts, and the End of Slavery* (Princeton: Princeton University Press, 2003), ch. 4.

159. On the impact and influence of women in the first millennium of the history of the church, see Mary T. Malone, *Women and Christianity. The First Thousand Years* (Ottawa: Novalis, 2000). The stories of Perpetua and Monica are recounted in Lynn Cohick and Amy Brown Hughes, *Christian Women in the Patristic World: Their Influence, Authority and Legacy in the Second through Fifth Centuries* (Grand Rapids, MI: Baker Academic, 2017), 27–64, 157–188.

Augustine's testimony about the role of his mother in his spiritual life is found in his *Confessions*, esp. Books I and III. On Augustine's pervasive influence on the way we see and understand ourselves in Europeanized cultures even until today, see Charles Taylor, *Sources of the Self: The Making of the Modern Identity* (Cambridge, MA: Harvard University Press, 1989), 127–142.

The broader, social influence of Christianity is summarized by Edmund Oliver, *The Social Achievements of the Christian Church*, reprint of 1930 ed. (Vancouver, BC: Regent College Publishing, 2004); Rodney Stark, *The Rise of Christianity: How the Obscure, Marginal Jesus Movement Became the Dominant Religious Force in the Western World in a Few Centuries*, (Princeton: Princeton University Press, 1996).

160. For an introduction to the role of Christianity in shaping modern conceptions of human rights, see John Witte Jr. and Frank S. Alexander, eds., *Christianity and Human Rights: An Introduction* (Cambridge: Cambridge University Press, 2010). For a more specific engagement with secular humanism, such as Rorty's, see Nicholas Wolterstorff, *Justice: Rights and Wrongs* (Princeton: Princeton University Press, 2010), and Michael J. Perry, *The Idea of Human Rights: Four Inquiries* (Oxford: Oxford University Press, 2000).

Chapter 7: Becoming Earthkeepers

161. Jurgen Moltmann, *The Trinity and the Kingdom: The Doctrine of God* (Minneapolis: Fortress, 1993), 9.

162. Michael Porter and Mark Kramer, "Creating Shared Value," *Harvard Business Review*, Jan–Feb 2011, available at: https://hbr.org/2011/01/the-big-idea-creating-shared-value (accessed Jan 20, 2015).

163. Eliyahu M. Goldrat and Jeff Cox, *The Goal: A Process of Ongoing Improvement*, 3rd Edition (New York: Routledge, 2016).

164. I've reflected briefly on Interface's story in Eduardo Sasso, "Sustainable Metaphors," *CSRWire*, February 3, 2017, available at: http://www.csrwire.com/blog/posts/1799-sustainable-metaphors (accessed Jan 21, 2018).

165. "The Responsible Company: Lessons from Patagonia's First 40 Years," Yale Center for Business and the Environment, September 27, 2012; webcast available at: http://cbey.research.yale.edu/events/the-responsible-company-lessons-from-patagonias-first-40-years (accessed Oct 12, 2012).

166. See David Miller, "Chick-fil-A and the Question of Faith in Business," *Harvard Business Review Blog Network*, August 9, 2012. It could be argued that owner and CEO Dan Cathy does this for marketing purposes, given that a high proportion of the population of the United States is (nominally) Christian and would presumably reward the company by purchasing on days other than Sundays. While there may be some truth to that, the policy is commendable as far as it seeks to be inspired by love of neighbor and not so much by considerations of economic value. (Does Cathy, after all, need the extra dollars? Not having to respond to unquenchable shareholder expectations every quarter, one is inclined to think that he doesn't.)

167. "Mayors of 7,400 Cities Vow to Meet Obama's Climate Commitments," *The Guardian*, June 28, 2017; see also: http://climatemayors.org (accessed Apr 20, 2018).

168. See Monica Araya, "A Small Country with Big Ideas to get rid of Fossil Fuels," TED, available at https://www.ted.com/talks/monica_araya_a_small_country_with_big_ideas_to_get_rid_of_fossil_fuels#t-123303; Bronwen Tucker, "Social Movements Played a Huge Part in Derailing Energy East," *CBC News*, October 12, 2017.

169. Statistics from the We Are Still In initiative, available at: https://www.wearestillin.com/faith-orgs (accessed Apr 17, 2018).

170. Ched Myers, ed., *Watershed Discipleship: Reinhabiting Bioregional Faith and Practice* (Eugene, OR: Cascade, 2016). Some inspiring stories and reflections are available at https://watersheddiscipleship.org (accessed Mar 20, 2018).

171. Cf. Neil Elliott, "Romans 13:1–17 in the Context of Imperial Propaganda," in *Paul and Empire: Religion and Power in Roman Imperial Society*, Richard Horsley, ed. (London: T&T Clark, 1997), ch. 11. See also chapter 5 above.

172. N. T. Wright, "Freedom and Framework, Spirit and Truth: Recovering Biblical Worship," in *Studia Liturgica* 2002, 32, 176–95. Building on the entire biblical prophetic tradition, one of the most explicit summons to speak truth into the lies of the beastly powers is found in Revelation 11:1–10.

173. See Cynthia D. Moe-Lobeda, *Resisting Structural Evil: Love as Ecological-Economic Vocation* (Minneapolis: Fortress, 2013), 271–300. Below I'm copying her own summary of the six gateways, all of which, as she rightly suggests, should be complementary (and inadequate if not pursued simultaneously alongside the others):

a. Small-scale business alternatives with emphasis on the local and regional, and on reducing consumption.

b. Moral culture within the business corporation.

c. Citizen action/consumer pressure to achieve "voluntary" constraints on corporate conduct.

d. Citizens using governments to achieve publicly mandated (regulatory or legislative) constraints on corporate conduct, and to limit the privatization and marketing of some essential goods (for example, water, seeds, and HIV/AIDS drugs).

e. Citizen action to rescind corporate personhood and the first amendment rights on the natural person that personhood grants to the corporation.

f. Organizing to expel or prohibit the establishment of unwanted corporations.

174. Martin Luther King Jr., *I Have a Dream: Writings and Speeches that Changed the World*, James Washington, ed. (San Francisco: HarperSanFrancisco, 1992), ch. 10.

Endnotes

175. King, *I Have a Dream*, 90.

176. Myers, *Watershed Discipleship,* offers a recent recollection of post-colonial stories reflecting reconciliation with several First Nations in North America.

177. The blend between dust and breath is one of the many valuable insights in Wendell Berry's aforementioned essay "Christianity and the Survival of Creation."

About the Author

A business sustainability consultant with a focus on spiritual ecology, Eduardo Sasso is a committed contributor to the renewable economy. In 2015, he co-founded Earthkeepers, an emergent citizen group in Vancouver aiming to live within a Judeo-Christian vision of ecology, public love, and climate justice. He holds degrees in engineering, theology, and sustainability from the University of Costa Rica, Regent College, and the University of British Columbia, respectively. Today he works and lives in Montréal, traditional territory of the Iroquois First Nation in the Saint-Laurent River Watershed.

www.ingramcontent.com/pod-product-compliance
Lightning Source LLC
Chambersburg PA
CBHW050810160426
43192CB00010B/1716